Aaron Fujimoto

CLASSICAL
GEOMETRY

CLASSICAL GEOMETRY

An Artistic Approach

Steve Pomerantz

Copyright © 2020 by Steve Pomerantz.

ISBN: Softcover   978-1-7960-8347-7
             eBook   978-1-7960-8346-0

All rights reserved. No part of this book may be reproduced or transmitted in any form or by any means, electronic or mechanical, including photocopying, recording, or by any information storage and retrieval system, without permission in writing from the copyright owner.

Any people depicted in stock imagery provided by Getty Images are models, and such images are being used for illustrative purposes only.
Certain stock imagery © Getty Images.

Four of the images used in the book were from The Metropolitan Museum of Art.

Print information available on the last page.

Rev. date: 01/30/2020

To order additional copies of this book, contact:
Xlibris
1-888-795-4274
www.Xlibris.com
Orders@Xlibris.com
804390

CONTENTS

Foreword ix
Introduction xiii

Part 1 Foundational Concepts for Classical Geometry

Chapter 1   Polygons 1
Chapter 2   Regular and Uniform Tilings.. 12
Chapter 2b   Square Tilings 18
Chapter 3   Tessellation 23
Chapter 4   Warm-Ups 28
Chapter 5   Overlay Tilings: The Polygon-in-Contact Method 34
Chapter 6   Variations on the PIC Method 38

Part 2 Further Examples and Advanced Grids

Chapter 7   Mamluk Grid.. 43
Chapter 8   Triangle-Hexagon-Triangle Grid 49
Chapter 9   Uniform Hexagonal Tiling and Trisectional Overlay 53
Chapter 10   Octagon-Square Grid 65
Chapter 11   Pentagon Grid... 77
Chapter 12   Dodecagon-Triangle-Square Grid 98
Chapter 13   3-3-4-3-4 Grid 107
Chapter 14   Dodecagon-Pentagon Grid 109
Chapter 15   The Rosette.. 115
Chapter 16   Dodecagon-Hexagon-Square Grid 128
Chapter 17   Topkapi Scroll... 134
Chapter 18   Jali Screen 140
Chapter 19   Dodecagon-Triangle Grid 145

Appendix 1   Even More Tilings... 151
Appendix 2   Completed Underlying Tilings 157
Glossary 163

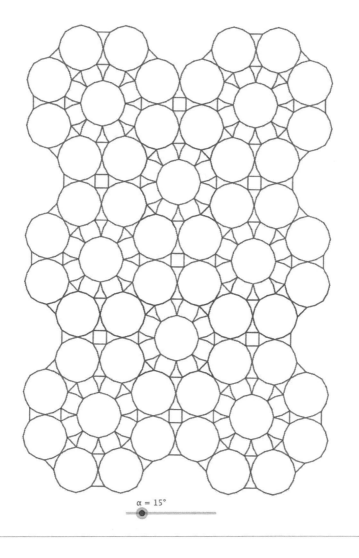

For me it remains an open question whether this work pertains to the realm of mathematics or to that of art.
—M. C. Escher

Foreword

It is my pleasure to introduce you to Steve Pomerantz's book *"Classical Geometry: An Artistic Perspective"*. I have watched his ideas about the intersection math and art grow from personal practice and theoretical discussion to what is now a strong, concrete resource for math educators (and learners) of all kinds.

We need this book, because we need teachers and students to learn that mathematics leads to beauty. At this moment in education history, "mathematics curriculum" consists of rote, compartmentalized concepts insofar as they apply to standardized tests. But historically, humanity developed math as a tool to engage in with the world around us. We sought to track time, build cities, and understand the heavens.

I am the director of the Monterey Bay Area Math Project. MBAMP is an organization whose mission is to increase the mathematical fluency of teachers and introduce math teachers to pedagogy that makes math instruction meaningful. Over the past six years, Steve collaborated with MBAMP, working directly with math teachers to enhance their pedagogy and subject knowledge—which we believe are deeply intertwined. In basic terms, Steve taught teachers to construct geometric patterns using a compass and a straight edge. As he led math educators in creating tilings, he made abstract concepts both real and playful. When you construct a hexagonal tiling, a perpendicular line is no longer just an idea, it is an important tool. When you construct a Malmuk grid, symmetry is not a bolded word in a text book, it is something personal and kinesthetic, that you created and care about. The patterns in this book spark conversation and curiosity around so many fundamental mathematical ideas and values—parallel lines, bisecting angles, polygons, uniformity, regularity, infinity, attending to precision, reflection, and more.

Every day students in K-12 math classes ask, "Why do we learn this?" Math came about so we could make sense of our world, using numbers, patterns, and shapes. It is a tool to relate to the complexity around us, and a reminder of its vastness. Students deserve math education that reflects this richness. *"Classical Geometry: An Artistic Perspective"* is a guidebook for engaging in mathematics in precisely this way. I look forward to using it in my own work with math educators of all grades, and I hope it opens all its readers to the beauty to mathematics.

Judith Montgomery
Administrative Director and Academic Facilitator of the
UCSC Math Project, MBAMP

Classical Geometry:
An Artistic Approach

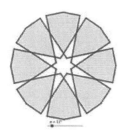

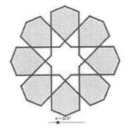

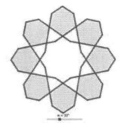

By: Steve Pomerantz

Introduction

Classical geometry lies at the heart of a thousand years of religious art and architecture. Stunning in their search of perfection, mosques, temples, and churches around the world achieve mathematical beauty with divine motivations.

On a more practical side, classical geometry represents an excellent application of the principles of Euclidean geometry. These religious compositions merge ruler-and-compass constructions, theorems discovered by the ancient Greeks, and fine craftsmanship to produce murals, paintings, and mosaics of infinite variety. This book explores classical geometry's intersection of art and mathematics. In the coming chapters, we will allow art to motivate a discussion of geometric fundamentals as the underlying math to enhance an appreciation of the art.

Most of the compositions in this text can be seen as patterns based on a tiling of the plane—and this is where the mathematical analysis begins. A *tiling* is a covering of the plane with polygons, which we will call individual tiles. In other words, it is an arrangement of shapes, such that they fill the plane without leaving any gaps. (In the coming pages, I use the terms tiling, grid, and pattern interchangeably.) As the book progresses, the individual tiles and larger tilings increase in complexity. The simplest individual tile is a regular polygon, a shape with equal sides and angles, such as a perfect square. The simplest tiling is a regular tiling, which is made up of identical, regular polygons. Think of a sheet of graph paper: the little and, of course, regular squares cover the page entirely, leaving no gaps.

For ease of reading, this book is divided in two halves. Part I lays out the fundamental concepts of classical geometry: the construction of polygons, the basic tilings made from those polygons, and secondary, overlay grids. Part II utilizes these concepts, teaching the reader to construct a series of increasingly complex patterns.

Most of the patterns I discuss in this book can be drawn directly on paper with just a compass and a pencil. As the patterns get more complicated, however, software may be helpful. I find Geogebra.org to be an excellent application. The Geogebra user will quickly discover many methods to draw the patterns presented here and discover their own new designs.

At the end of the book, you will find an appendix with completed grids to play with and perhaps overlay. For the reader still intrigued after finishing all the following patterns, I would refer you to Jay Bonner's wonderful book *Islamic Geometric Patterns*, which offers an endless variety of patterns, steeped in history, for your education and amusement.

Many more wonderful resources are available. For starters, I recommend the Prince's School of Traditional Art and the Art of Islamic Pattern; both London-based schools with on-line resources, that have been very inspirational to me.

I would also like to thank many who have helped me along the way including the Monterey Bay Area Math Project, the Creative Geometry Salon at the Anthroposophy Center of New York, Anna Montgomery, countless teachers and Maggie for her endless patience and support.

Part I

Foundational Concepts for Classical Geometry

This book serves as a practical manual for the construction and understanding of classical geometry patterns. Generally, these tilings are created in two stages. First, an initial grid is constructed out of polygons, and then a secondary tiling is overlayed using a method called Polygon-in-Contact (PIC). Once that overlay pattern (also called a dual or secondary tiling) is drawn, the underlying grid is typically erased, leaving only the final design. Part I begins with the construction of individual polygons and culminates in an introduction to the PIC method and some simple overlay grids.

Chapter 1

Polygons

The essential building block of tiling is the construction of polygons. The simplest polygon, a *regular polygon*, has sides of equal length and vertices of equal angles. For a regular polygon, an interior angle measures $180 \times (1 - 2/p)$, where p is the number of sides. For example, for a square, $p = 4$, and the formula above gives $180 \times (1 - 2/4) = 90$. There are as many regular polygons as there are integers for there to be sides.

The basic polygons will introduce us to basic construction methods: triangles (3 sides), squares (4), pentagons (5), hexagons (6), octagons (8), decagons (10) and dodecagons (12). We will begin in the middle, with the construction of a hexagon. First, draw three circles with equal radii, all centered on a line. The two outer circles are centered on the points where the inner circle hits the horizontal line (B, D). The four intersection points of the outer circles with the inner one (E, G, H and F), along with points B and D, mark the six points of a hexagon.

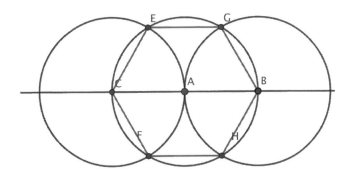

Figure 1.1 The construction of a hexagon.

1

Next, to draw a square, we will continue using the circle centered at *A*. First, create a second line at point *A*, perpendicular to the horizontal one. To do this, draw equal-radii circles about *E*, *G*, *F*, and *H*. It is not necessary to draw the entire circle, just draw enough to identify where the circles intersect, which we will label points *L* and *N*. The line through $\overline{LN}$ is now perpendicular to the original horizontal line.

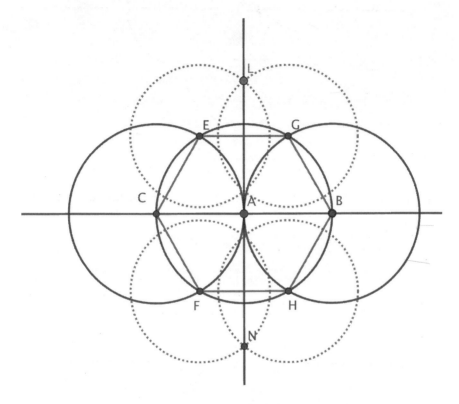

Figure 1.2

Now, draw circles above and below this horizontal centered at *P* and *Q* that have the same radii as the original circles about *B* and *D*. These four circles intersect each other at *C*1, *U*, *W*, and *A*1, as labeled below.

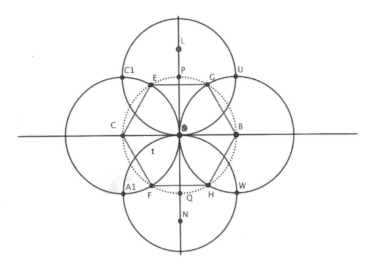

Figure 1.3

Finally, by connecting these four points, we have a square that is outside the original circle.

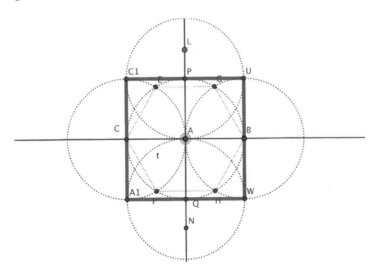

Figure 1.4

Alternatively, we can construct a square by identifying where the diagonal lines through $\overline{C1W}$ and through $\overline{A1U}$ cross the original circle. Connect the four points R, M, J, and T to create a square inside the circle.

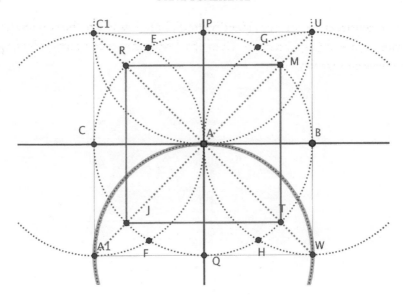

Figure 1.5

Another square, rotated relative to the first can also be inscribed inside the circle (*P*, *B*, *Q*, *C*). Looking like a baseball diamond, the additional square can be found inside the same circle by joining the top, bottom, and two sides of the circle. The diagram below illustrates both squares.

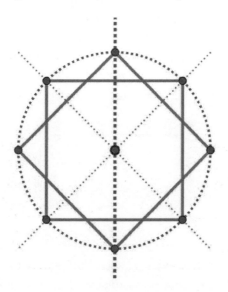

Figure 1.6

To construct an octagon, identify the eight points where the two interior squares intersect (e.g., point *A*). Connected with the seven others points, we now have an octagon.

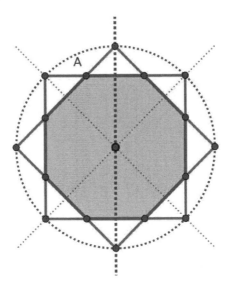

Figure 1.7

An alternative construction of an octagon, begins with a simple square. Identify the center of the square and measure the distance from any corner to this center. Using an arc, find eight points around the square perimeter that are of equal distance from the corners. Connect these points to produce another octagon.

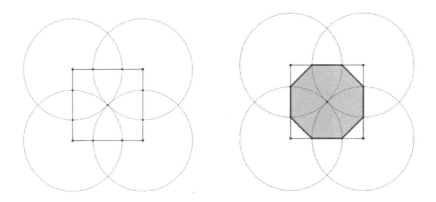

Figure 1.8

To construct a dodecagon, return to our construction of the square. At the step when there were four circles surrounding the original circle, make note of their points of intersection with the central original. We find the twelve points of the dodecagon.

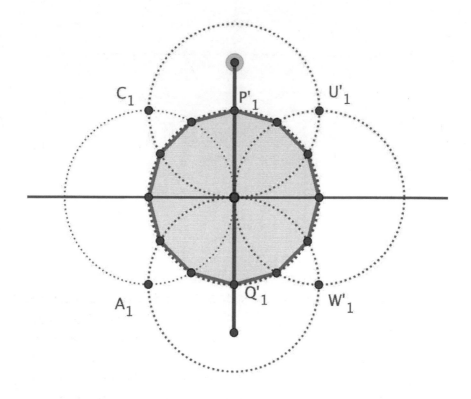

Figure 1.9

Constructing the pentagon is perhaps one of the more complicated. The construction illustrated differs from Euclid's approach but similarly relies on the golden mean.

We begin with a defining circle (A) and two smaller circles (E and G). G is drawn first, centered at the midpoint of the horizontal segment $\overline{AD}$ with radius equal to this mid-length. E is then drawn to be tangent to circle G at H.

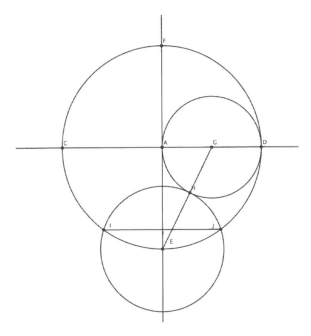

Figure 1.10

Now, if $\overline{AE} = 1$, then $\overline{AG} = \overline{GH} = 1/2$, and thus we calculate $\overline{EG} = \sqrt{5}/2$. We also know that $\overline{EJ}$ is a radius of E. In other words, $EJ = EG - GH = \sqrt{5}/2 - 1/2$. Note that $\angle FJE$ is a right angle, so we calculate that $\sin(EFJ) = \overline{EJ}/\overline{FE} = (\sqrt{5} - 1)/4$, which is exactly $\sin(18°)$, so $EFJ = 18°$.[1] From this, we obtain that $\angle EAJ = 36°$ because it has the same arc but has the center of the circle as the vertex. Then we can calculate $\angle IAJ$ to be $72°$, which is $1/5$ of the circles $360°$. This establishes $\overline{IJ}$ as a side of the inscribed pentagon. The complete construction appears below.

---

[1] This follows from the clever use of half-angle and double-angle formulas and the convenient observation that $90° = 72° + 18° = 2 \times 36° + 36°/2$. Then the identity $\cos(90°) = 0$ provides a quadratic equation for $\cos(36°) = 0$, from which the double angle formula provides $\sin(18°)$.

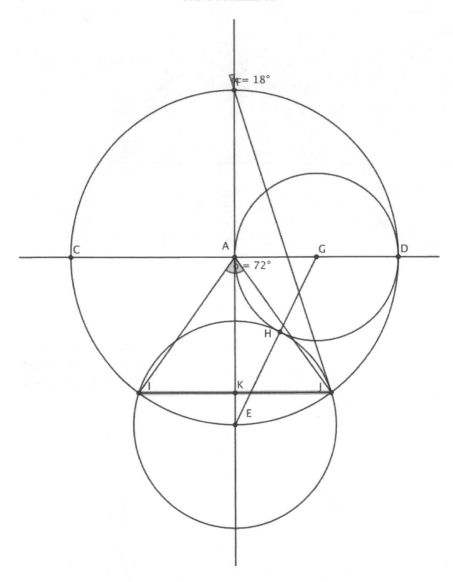

Figure 1.11

Once $\overline{IJ}$ has been identified, it can be copied around the circumference to obtain the pentagon.

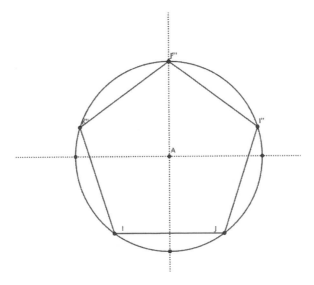

Figure 1.12

Once we have a polygon with a given number of sides, it is easy to divide those sides in two, thus doubling the number. For example, once we have a pentagon, we can draw a diagonal connecting a vertex with the center and extending it until it reaches the other side.

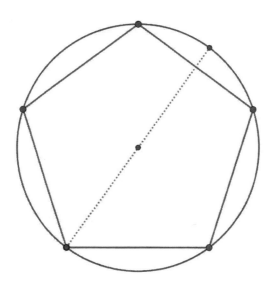

Figure 1.13

Repeating this, we can identify all the points for a decagon.

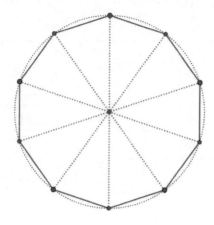

Figure 1.14

However, when the original polygon has an even number of vertices, this technique will not work. In order to determine the other vertices, we need to bisect the central angle that contains two adjacent points as follows.

Consider an octagon that identifies eight points on the circle, and suppose we want to identify sixteen. Construct two circles about adjacent vertices, and identify where they intersect. The radius is not important, only that it be large enough for the circles to intersect.

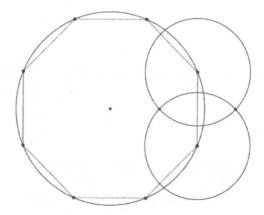

Figure 1.15

These two intersection points define a line that bisects the side of the octagon and identify a point on the circle midway between the two vertices.

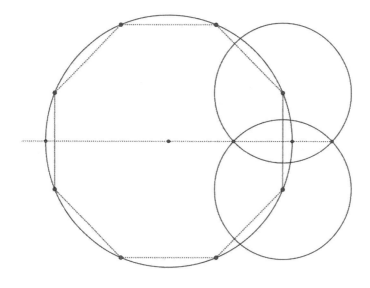

Figure 1.16

This new point defines the length of the new 16-sided figure.

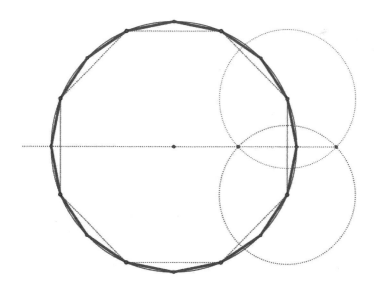

Figure 1.17

Chapter 2
Regular and Uniform Tilings

The simplest tilings, *regular tilings*, are made out of a single, tessellated (repeated) regular polygon. There are only three possible regular tilings. If all the polygons that meet at a vertex are the same, then the sum of the interior angles of the polygons that meet at that point must add up to 360, and there are only three ways to achieve this: triangles ($120 \times 3 = 360$), squares ($90 \times 4 = 360$), and hexagons ($60 \times 6 = 360$).

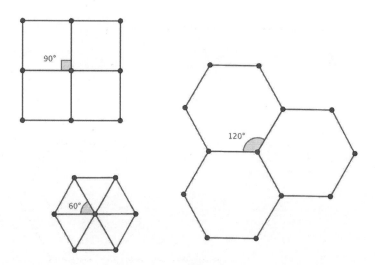

Figure 2.1: The three possible regular tilings.

If we allow the individual polygonal tiles to vary but stay regular, there are eight more possible polygon tilings that will cover the plane. These are called *uniform tilings*.

Two of these tilings are made up of two squares and three triangles. In this case, the angles are $90 + 90 + 60 + 60 + 60 = 360$.

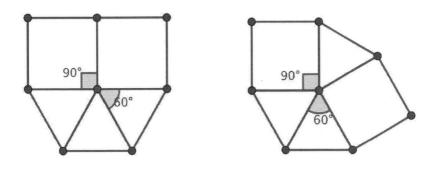

Figure 2.2

The third uniform tiling is constructed with two hexagons and two triangles.

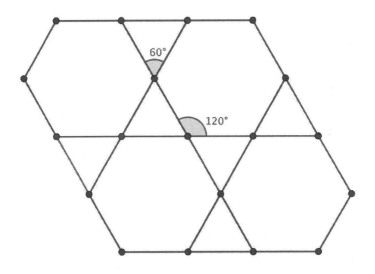

Figure 2.3

The fourth has two octagons and a square at each point.

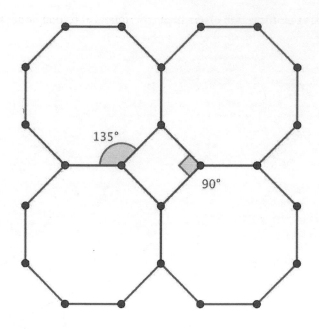

Figure 2.4

The fifth is made of two dodecagons and a triangle.

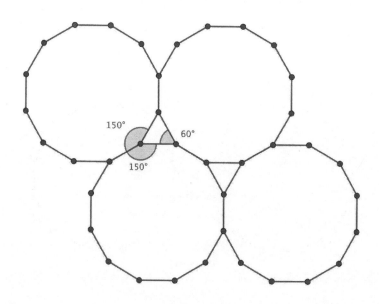

Figure 2.5

The sixth uniform tiling has a hexagon, square, and triangle at each point.

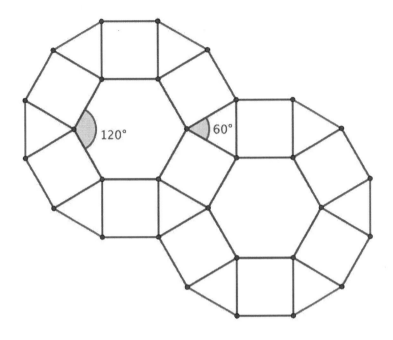

Figure 2.6

The seventh has a dodecagon, hexagon, and square.

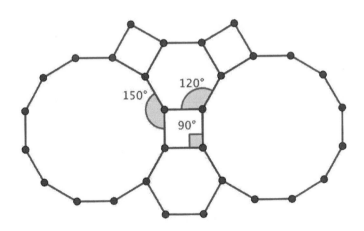

Figure 2.7

And finally, the eighth uniform tiling has a hexagon and three triangles at each point.

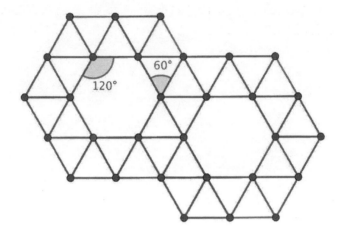

Figure 2.8

There are additional tilings of varying regular polygons that fit together at a point because the sum of their interior angles add to 360, but these cannot be extended to fill the plane.

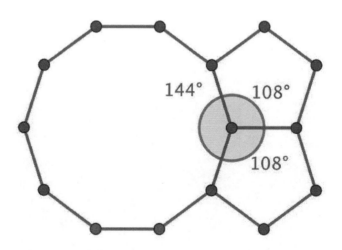

Figure 2.9: These regular polygons fit at a point but cannot be extended to fill the plane.

The figure above is based on regular polygons, pentagons and decagons. The angles add up to 360 at the point circled in green, but the pattern cannot be repeated to fill a plane without gaps. If a polygon has p sides, then each interior angle has $(1-2/p) \times 180°$. There are other examples like the one above. Here's a hint: (42,7,3), (24,8,3), (18,9,3), (15,10,3), (20,5,4), and (10,5,5) as shown above.

Chapter 2b
Square Tilings

We now begin a new pattern by bisecting one side of a square and centering a circle at the midpoint E as illustrated.

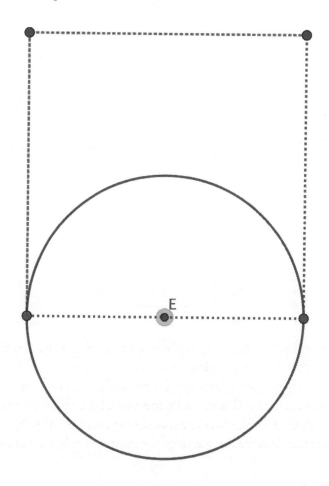

Next, create a line segment from the corner at an arbitrary angle, here illustrated to be 60.

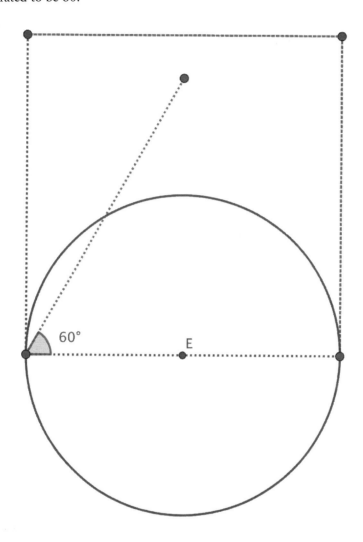

If G is the point where the ray intersects the semicircle, and then G is connected to the lower right corner, we create a right angle at G because any triangle with a hypotenuse given by a diameter of a circle is a right triangle. If the ray is chosen so that G is at the intersection of the circles about A and E, then the angle GAE will be 60. Note that the hypotenuse of this right triangle, being a diameter, is twice the length of the shorter side, being a radius.

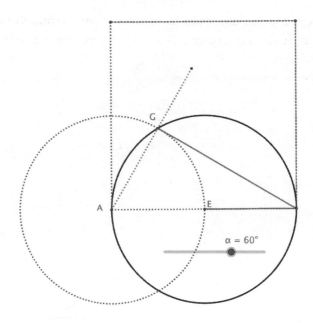

Repeating this construction on each side of the square produces the pattern shown below. This square, made up of a smaller square with four triangles, is our individual tiling unit to be tessellated.

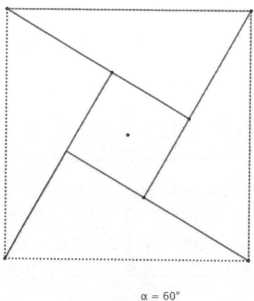

Then by repeating this piece in horizontal and vertical directions, we can obtain the following pattern. Thus, we can tessellate the plane with this fundamental unit.

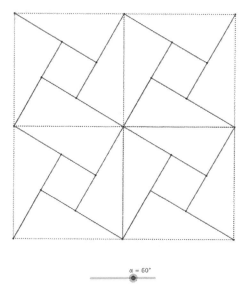

A more traditional pattern, from the Fatehpur Sikri in India, is created if, when tessellating, we reflect the individual squares.

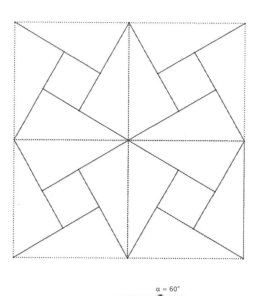

Some larger grids:

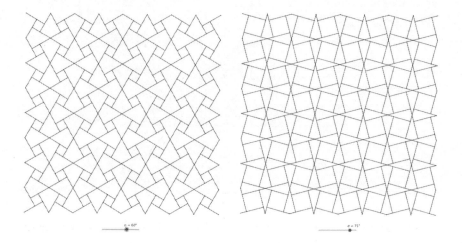

Chapter 3
Tessellation

In the previous two chapters, we learned to construct polygons and then established that certain combinations of polygons can cover a plane. Now, we will continue on to create a completed tiling. In this chapter, we construct individual composite tiles that, when combined, *tessellate* the plane. A tessellation is a pattern that, when repeated, covers the plane, with no gaps or overlaps.

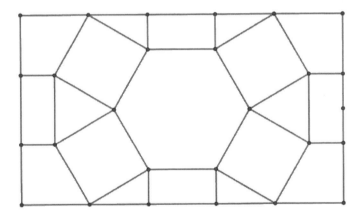

Figure 3.1 A 6-4-3 tile, ready to be tessellated and fill the plane.

Let's begin by constructing a "6-4-3 tiling," made up of hexagons, squares, and triangles. In order to tessellate the plane, we aim to produce a rectangular section that looks like Figure 3.1 above. This is made up of the 6-4-3 pieces, but as a rectangle, it can easily be repeated to fill a region. In fact, any rectangle, not just squares, will tessellate the plane. If we can construct a rectangle based on polygons, then it can be used to fill an entire plane, surface, or wall.

The construction begins with a circle centered on a horizontal line and two other circles drawn on either side. These additional circles have the same radius and are centered on the original circle. The brown rectangle, cornered at the intersection of these circles, will be the rectangular unit to tessellate.

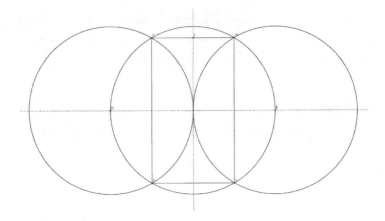

Figure 3.2

Find the midpoint of the horizontal side, J, and draw a circle around it that is tangent to the two horizontal sides. We can now construct a smaller circle, centered in the middle of the rectangle. The radius of this inner circle is chosen so that it is tangent to the larger circle above it.

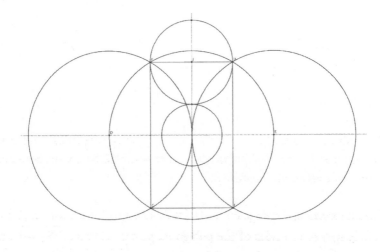

Figure 3.3

Placing a circle above and below this middle circle will be helpful for further constructions. The central circle now holds the hexagon for the final pattern. The hexagon can be drawn by marking off the radius of the circle along the circumference beginning at the top point, *K*.

In the second picture, a similarly sized circle is drawn at the rectangle's corner. Points *U*, *W*, and *Q* are found at the intersections of the drawn circles and form a square.

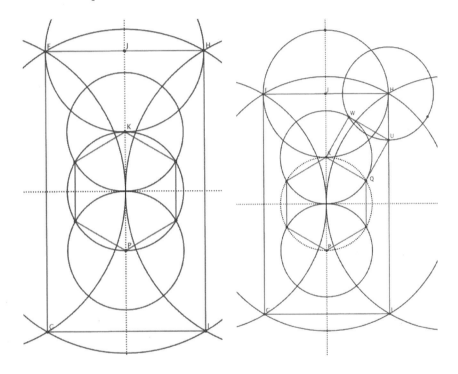

Figure 3.4

With these lines drawn in so far, we can now see why the inner circle will fit this pattern. The top half of the rectangle is shown below in more detail with some of the lengths identified.

In this example, the large radius $OH = 2$. The smaller radius, which I call r, is the length of the sides of the polygonal pieces, example $\overline{OK}$, $\overline{KW}$, and $\overline{WH}$. Now, $\overline{OH}$ is also made up of three segments: $\overline{OKQ}$, $\overline{KQWU}$, and $\overline{WUH}$.

The middle segment $\overline{KQWU}$ is just r, but the other two segments are altitudes of an equilateral triangle of side r, each of which is $r \times \cos(30°)$.

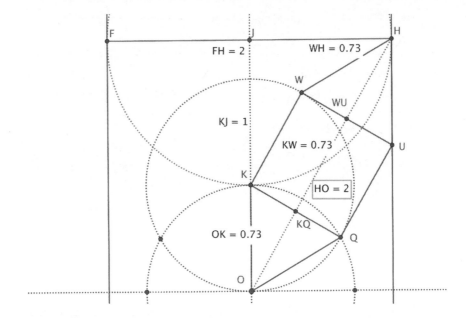

Figure 3.5

We see that $2 = r \times \cos(30°) + r + r \times \cos(30°) = r(1 + \sqrt{3})$, from which $r = 2/(1 + \sqrt{3}) = \sqrt{3} - 1$ (approximately equaling 0.73).

Now, this value can be found exactly because $\overline{OJ} = \sqrt{3}$ and $\overline{JK} = 1$. The inner circle is thus found by "subtracting" $\overline{JK}$ from $\overline{OJ}$.

Continuing to the other corners, a circle drawn identifies the remaining squares in the pattern. Triangles above and below the central hexagon can now be completed as well. A pair of horizontal and vertical lines now identify the rest of the pattern.

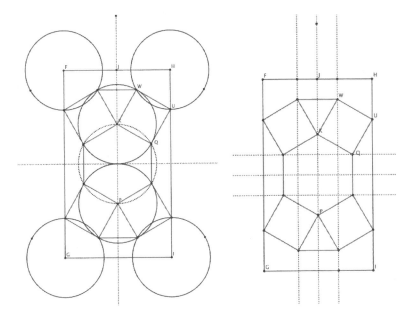

Figure 3.6

The final pattern with guidelines removed appears below. As intended, this pattern can be tessellated to create a larger tiling. These tilings are where classical geometry designs begin. In some traditions, such as early Roman Mosaic Art and the Cosmatesque tradition, these designs form the basis for wonderful art. Most Islamic traditions, however, use the above tilings as a jumping-off point for more complex designs as we will see in the following chapter.

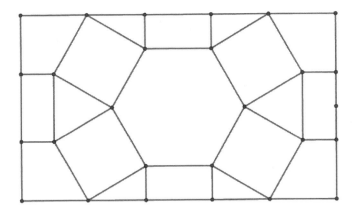

Figure 3.7

Chapter 4
Warm-Ups

Before delving into the main technique for drawing, I introduce two significant designs because of their very common usage.

The first will be a ten-fold rosette fit inside a rhombus. The rhombus shape can tessellate the whole plane, so this is a useful design.

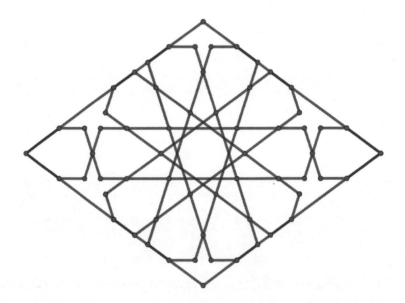

We begin with the decagon and draw a star inside by connecting every third point.

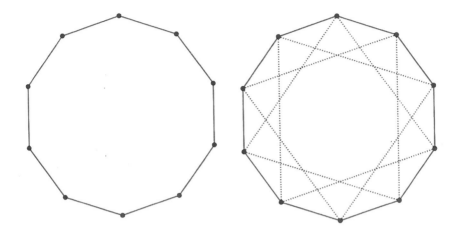

Now, first, identify the points of self-intersection of this inner star, and draw a pair of parallel lines through them, stopping at the edge of the decagon. The first pair is illustrated below left, and then continue by drawing a total of 5 pairs of parallel lines.

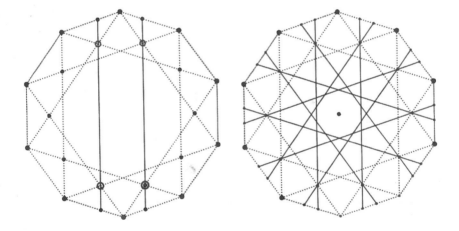

The tops of the 10 petals are now found as segments of the outer decagon, which completes the rosette.

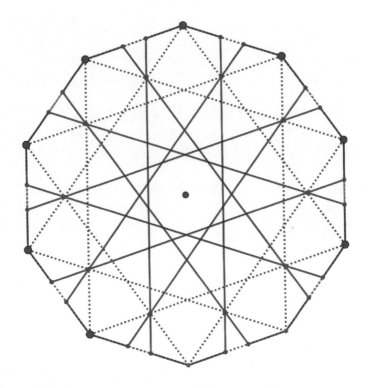

To form the repeating rhombus unit, first extend 4 of the decagon sides to identify the edge. Then extend the top and bottom petal and the two of the original parallel lines to reveal the remainder of the design.

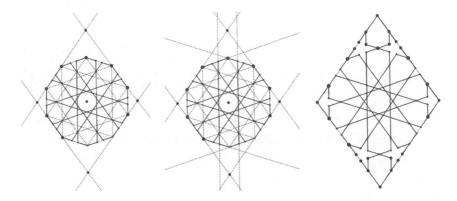

The rhombus can then be tessellated to fill an arbitrary region.

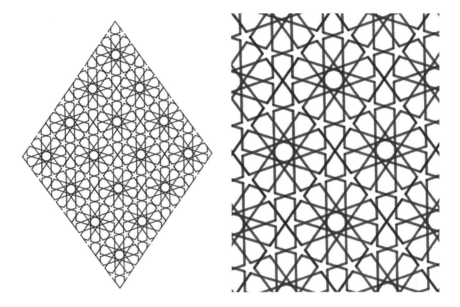

The next pattern to discuss is based on the Khazan Tiles, sometimes called Star and Cross.

We begin with horizontal and vertical lines through a circle and then similar size circles centered at the 4 nodes of this center circle.

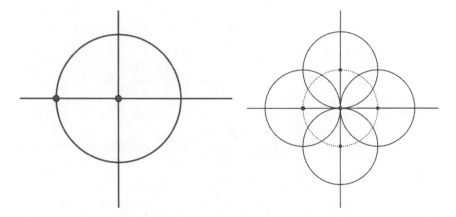

The intersection of these 4 circles allows for the construction of 2 diagonal lines through the center, which now identifies 8 equally-spaced points around the center circle. Connecting these points with 2 squares, one normally oriented, the other like a diamond, outlines the shape of an 8-pointed square.

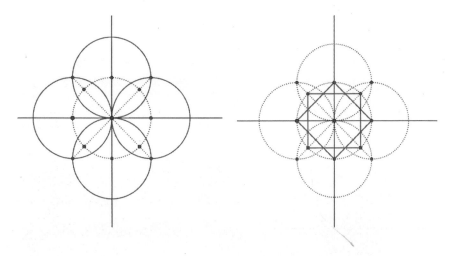

Classical Geometry

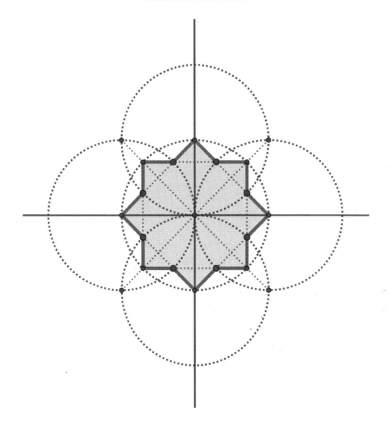

By forming new circles along the axes, the pattern can now be extended.

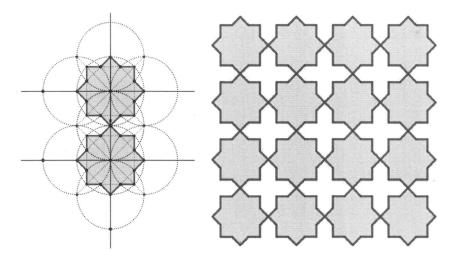

Chapter 5

Overlay Tilings: The Polygon-in-Contact Method

In most classical geometry, the final tiling is obtained from a secondary grid overlayed on top of an initial construction. These dual or overlay tilings can be obtained quickly using the PIC method. Through simply connecting the sides of an underlying pattern's polygons, the PIC method allows us to extend these patterns to endless variation and complexity. It is particularly in the construction of dual tilings that the use of software is most helpful.

The PIC method proceeds by finding the midpoint of each polygon side in the underlying grid. For example, in Figure 4.1 below, I located the midpoints of all sides of the triangles, squares, and hexagons.

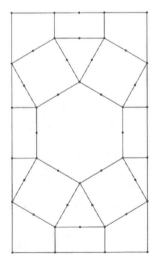

Figure 4.1: A 6-4-3 tiling with midpoints marked out.

Next, choose an angle from which you will to introduce the overlaying grid. The angle is somewhat arbitrary. By varying the angles, we can produce different patterns. As we will see, there are certain angles that give rise to traditional or aesthetic patterns, but any angle can be chosen, and I encourage you to experiment. In Figure 4.2 below, I have chosen a 55° angle.

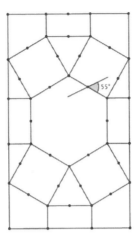

Figure 4.2: A 55° angle has been selected.

Next, draw a line at each midpoint across the polygonal side at the chosen angle. Terminate the individual segments when they intersect another newly segment coming off the midpoint of an adjacent side.

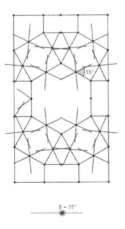

Figure 4.3

You will produce a new pattern that overlays the original grid. Below, in Figure 4.4, you can see the original grid lines, and the overlay pattern is drawn as solid tiles.

Figure 4.4

Finally, remove the underlying grid to complete the tiling.

Figure 4.5

CLASSICAL GEOMETRY 37

A wide range of angle choices produce a wide range of patterns:

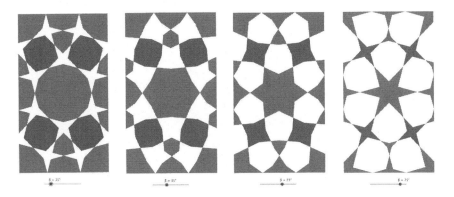

Figure 4.6

Chapter 6
Variations on the PIC Method

Once the polygonal grid is established, the overlay design need not be based on the midpoints of the sides. In this example, line segments originate from the vertices and continue until they intersect each other.

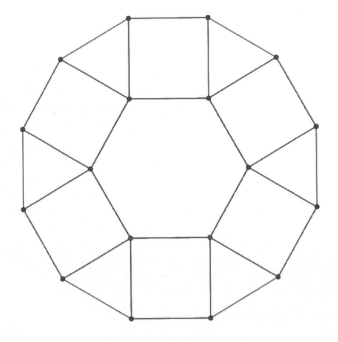

Figure 18.1

A classic example is based on a 30° angle, which converts the triangles to hexagons.

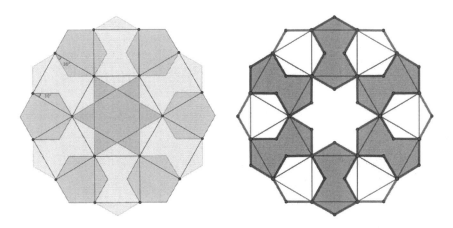

Figure 18.2

This large hexagon can now be tessellated to fill larger regions.

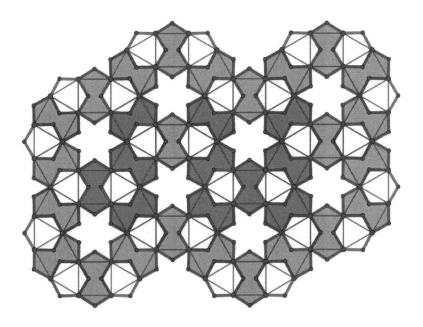

Figure 18.3

Choosing different angles produce different patterns; the 45° is illustrated below.

Figure 18.4

An alternative is referred to as a two-point pattern. In this approach, two points are identified on each edge rather than the one midpoint. This method is discussed in more detail in the discussion on hexagons.

Part II

Further Examples and Advanced Grids

In the next chapters, you will find a series of more complex grids, overlays, and techniques. Part II does not proceed sequentially as the previous section did but, rather, for the most part, contains a series of stand-alone examples.

Chapter 7
Mamluk Grid

For our next design, the Mamluk Grid, we will create a tile to tessellate, but the fundamental unit will be hexagonal. This pattern begins with the underlying hexagon-square-triangle grid constructed in a new way. Start with a horizontal line, a center circle, and proceed to draw six equal-radii circles around the circumference of this central circle.

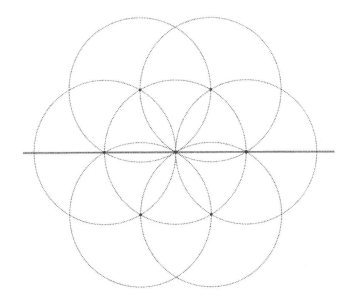

Figure 5.1

The intersection of these outer circles with the central circle forms the inner hexagon. Through opposing points of this hexagon, draw three sets of parallel lines, and note where they intersect these outer circles.

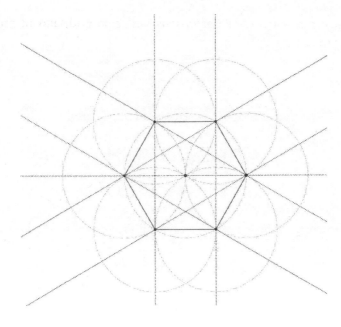

Figure 5.2

These intersection points will be the vertices of the squares and triangles.

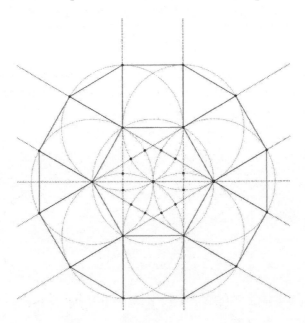

Figure 5.3

Now, begin with this hexagon-square-triangle grid, and identify the centers of the triangles.

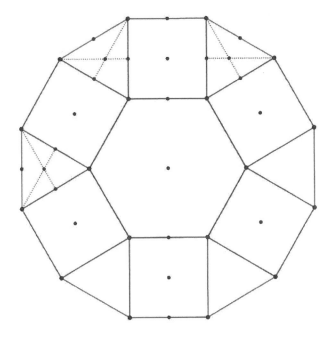

Figure 5.4

This allows for the determination of an inscribed circle. Then identify six points on the circumference of this circle and draw three sets of parallel lines.

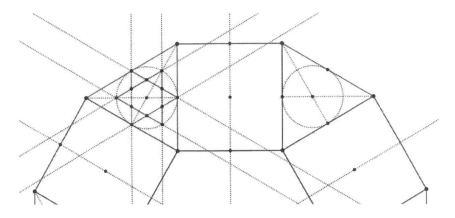

Figure 5.5

Extend the lines from the central six-pointed star to the radii of the larger pattern as illustrated below.

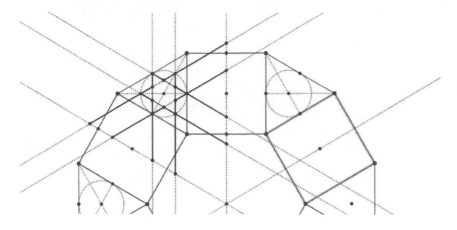

Figure 5.6

Repeat this construction at each triangle by finding the inscribed circle and constructing three pairs of parallel segments. The inscribed circle can be found by identifying the intersection of the angle bisectors.

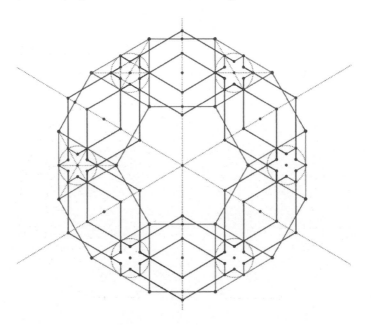

Figure 5.7

Additional interior lines are drawn by first drawing a smaller rhombus that has dimensions identical to one of the points of the six-pointed star. Connecting the corners of these six rhombi gives us a new hexagon.

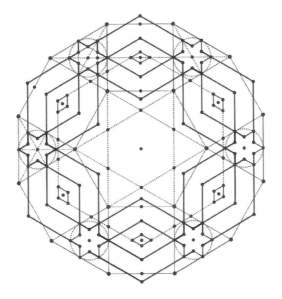

Figure 5.8

Remove the auxiliary lines, focusing on the interior of the red hexagon, to reveal the tiling unit. Tessellate to create a final grid.

Figure 5.9

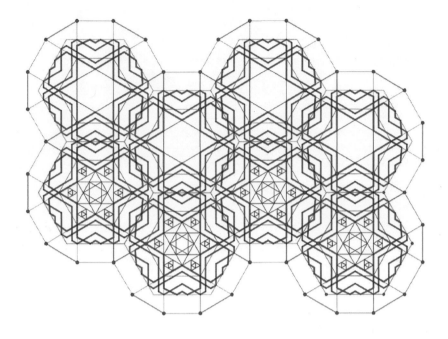

Figure 5.10

Chapter 8
Triangle-Hexagon-Triangle Grid

Similar to other hexagon grids, this grid begins with a horizontal line and three circles. The intersection of the circles provides the vertices of the initial hexagon.

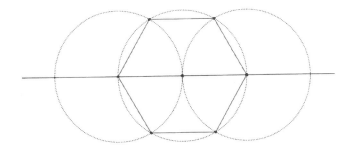

Figure 6.1

By adding two more circles to the right side and identifying the intersections, you can identify a second hexagon in the same manner as above. Add two short horizontal lines to create the triangles in between them.

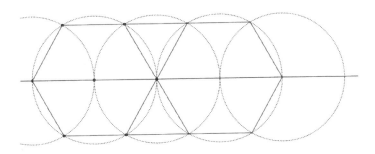

Figure 6.2

Now, add more circles below to reveal a third hexagon below the lower triangle.

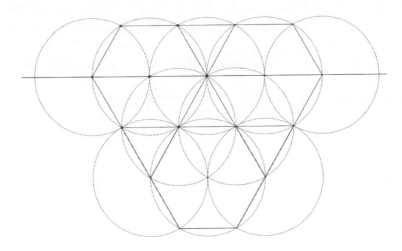

Figure 6.3

Continue adding circles to create hexagons and triangles until a complete grid is created. This will be our underlying grid.

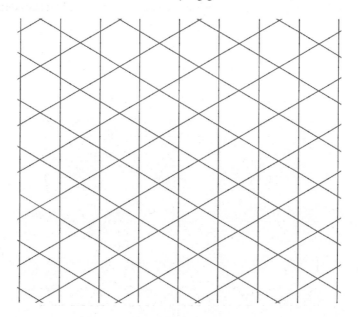

Figure 6.4

Next, create the overlay grid. As before, begin by marking the midpoints of each side. Then draw a line segment that passes through each midpoint at the angle of your choice. Below, you can see a dual tiling of this grid with a 45° angle. Terminate the individual segments at the point where they intersect another newly drawn segment coming from an adjacent side. For example, see the dual tiling of this grid with a 45° angle shown below. (For extended directions, see Chapter TK.)

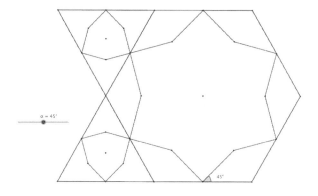

Figure 6.5

Apply this method to the entire grid to produce a pattern of linked dodecagons that appear to replace the six-pointed stars inside the original hexagon.

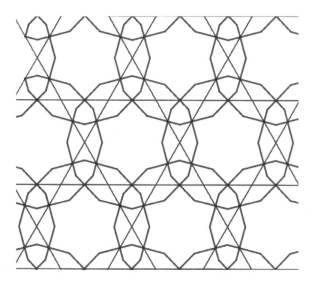

Figure 6.6

If we select a 30° angle, it produces a pattern of large hexagons ringed by small hexagons:

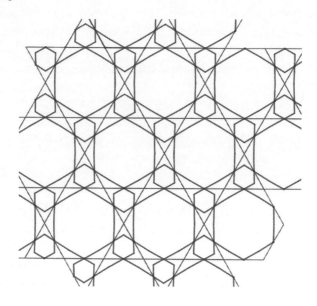

Figure 6.7

With a 75° angle, we obtain large and small pointed stars. Or are they dancing squares?

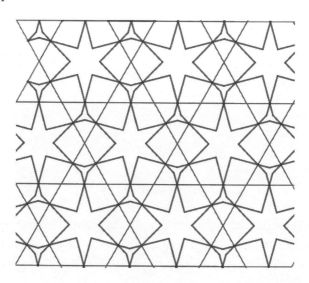

Figure 6.8

Chapter 9

Uniform Hexagonal Tiling and Trisectional Overlay

A uniform hexagonal tiling is one of the most straightforward tilings to construct. Begin with a horizontal line. Construct a circle around point *A*, centered on the line, then two circles on either side, with the same radius as circle *A*, and centered on the intersection points of circle *A* and the horizontal line. Mark the points where these outer circles intersect the center circle (e.g., Point *E*).

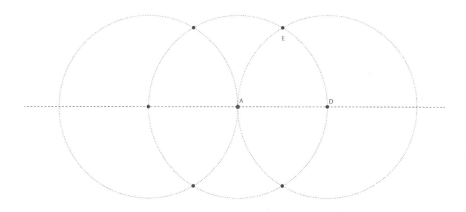

Figure 7.1

Since these circles have the same radius, the triangle connecting points *A*, *E*, and *D* is equilateral, hence equiangular. This insures that by connecting the six points identified above, we have constructed a hexagon with an interior angle equal to 120°.

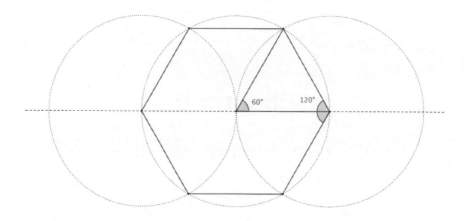

Figure 7.2

Now, continue to draw more circles with centers at the intersection of previously drawn circles as shown below.

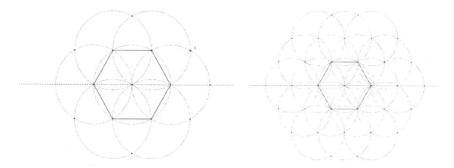

Figure 7.3

Inside each of these circles, a hexagon can be drawn with vertices given by the points where the circles intersect.

Classical Geometry

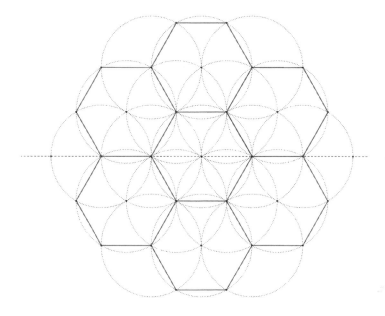

Figure 7.4

Remove the circles to yield the underling grid.

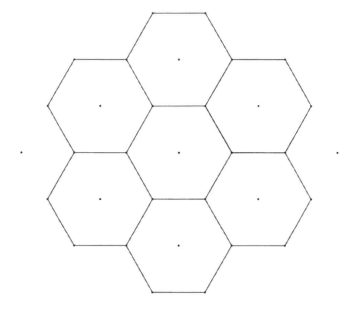

Figure 7.5

Mark the midpoints of the sides to start the overlay design.

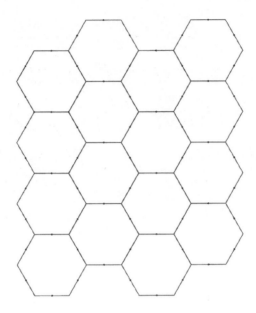

Here, as an example, I selected a **70°** angle for the overlaying line segments, which forms a star shape.

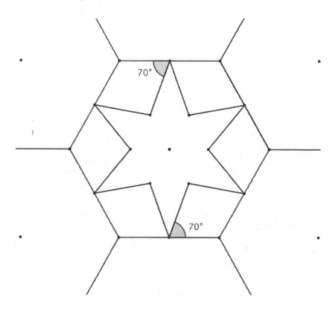

Figure 7.6

Tessellate the star to fill a region.

Figure 7.7

Below a smaller, 45° angle produces a more rounded shape.

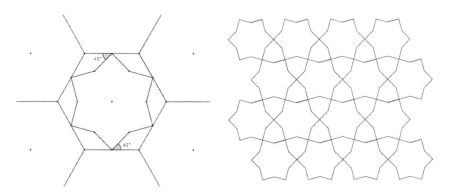

Figure 7.8   Figure 7.9

A 60° angle produces this very traditional pattern.

(MMA – Iran 13th Century)

Figure 7.10

Using different angles produces different designs. This is an appealing feature of the PIC method. Once the underlying grid and midpoints are identified, the choice of angle used to create the overlay design is arbitrary.

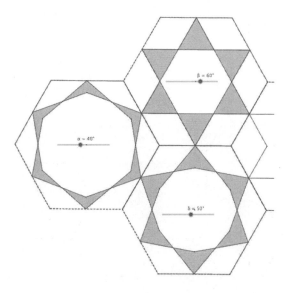

Figure 7.1

For an alternative overlay design, trisect each side, i.e., divide each side into three equal lengths.

A line segment that is part of the hexagonal grid can be trisected by the following technique: Draw a diagonal line that connects the opposing endpoints of two vertical sides. The intersection points of that line and the intermediate vertical line segments produce exact one-third divisions.

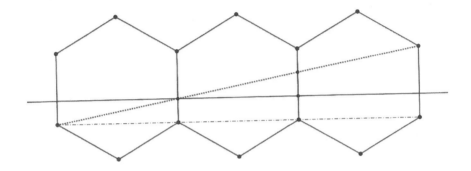

Figure 7.12

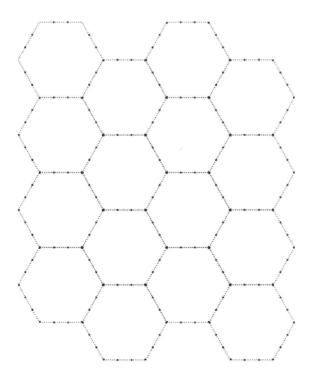

This method is referred to as a two-point design. Then draw segments from each of these points at the angle of your choice. For example, below, I selected a 45° angle.

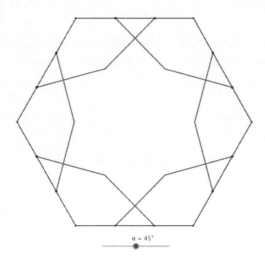

Figure 7.13

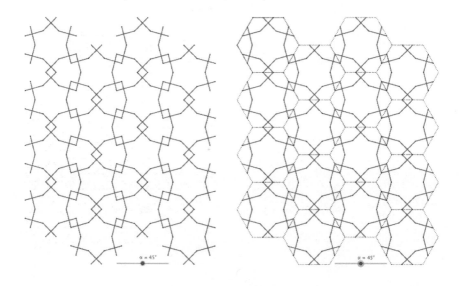

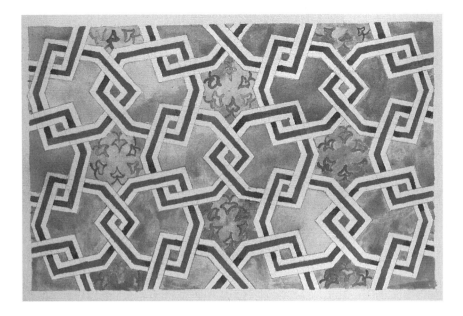

Figure 7.14

Next, we see a grid with 60° angles selected.

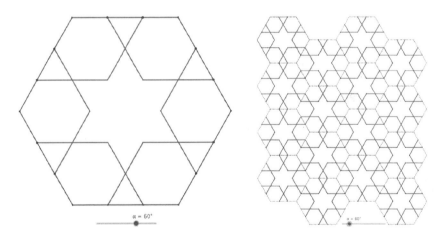

Figure 7.15

The trisection of the hexagon side allows us to elaborate on the above and construct an interior star as shown:

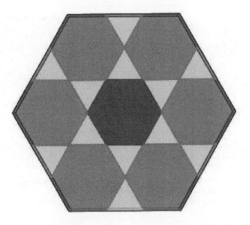

Figure 7.16

Proceeding from the larger grid of hexagons, draw lines connecting the opposing corners as illustrated below. This is another method for trisecting sides—but only of the interior hexagon.

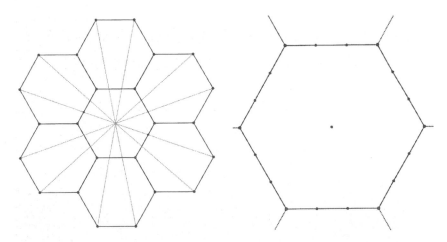

Figure 7.17

Now, connect those points to form three sets of parallel segments and produce the star.

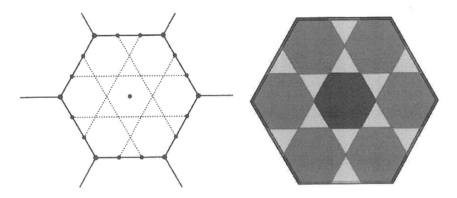

Figure 7.18

Tessellate the hexagon to produce the following grid:

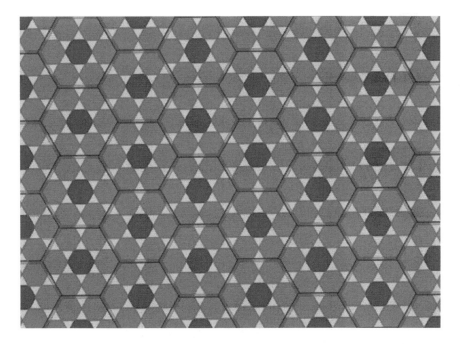

Figure 7.19

While on the topic of hexagons, a particularly simple example of an alternative underlying grid is illustrated below. In this case, the hexagons are aligned differently than in the regular tiling and leave room for rhombi in between.

Continuing with the PIC method generates the designs below. The first picture shows a 60° overlay, followed by 45°, 72°, and 30°.

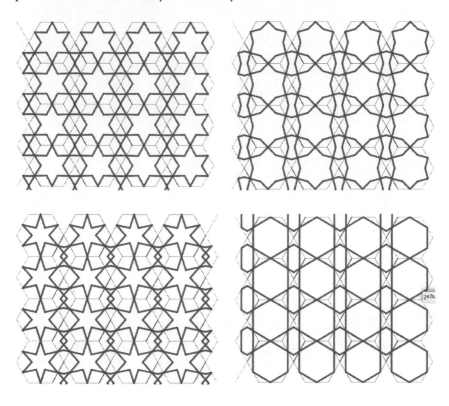

Figure 7.20

Chapter 10
Octagon-Square Grid

The next construction is a tiling made of octagons and squares, which can form the underlying grid for many other types of designs, such as the one below.

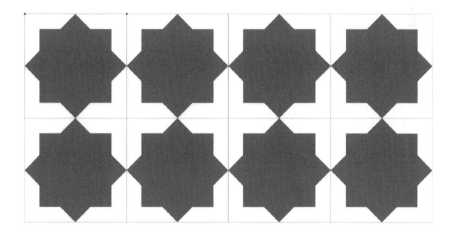

Figure 8.1

To begin, we will construct an octagon by drawing a square. Identify the center of the square and drawing a circle centered at each corner with a radius set as the distance between a square's corner and its center.

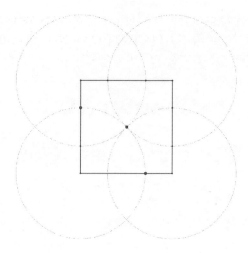

Figure 8.2

The intersection points of these circles with the edges of the square form the corners of an octagon, as illustrated below. To understand this, imagine that the side $\overline{AB}$, in the drawing below, has length 2. Then the radius of the circles drawn is $\sqrt{2}$. Therefore, $\overline{BG} = 2 - \sqrt{2}$ and $\overline{GH} = 2(\sqrt{2} - 1)$. Now, $\overline{GJ}$ is the hypotenuse of an isosceles right triangle with side $\overline{BG}$ and has length $\sqrt{2} \times (2 - \sqrt{2}) = 2(\sqrt{2} - 1) = \overline{GH}$.

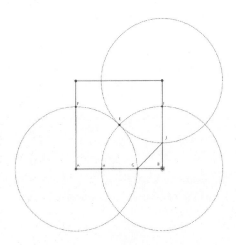

Figure 8.3

Connecting these edge points all around produces the octagon. For now, we will not construct the squares of this square-octagon tiling from the corners, but V*JGB* will be one quarter of one of the squares.

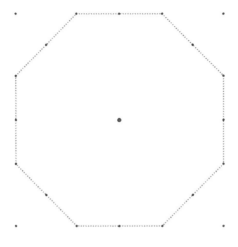

Figure 8.4

A traditional pattern can now be constructed by drawing line segments from the midpoints at 45° angles. This creates two overlapping squares inside the octagon: one normally aligned and another at a diagonal. Drawn below is the octagon filled in this way and tessellated.

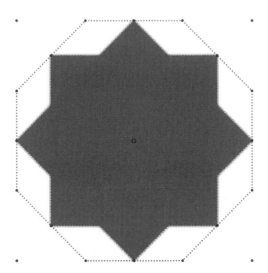

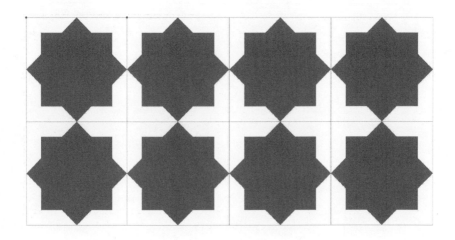

Figure 8.4

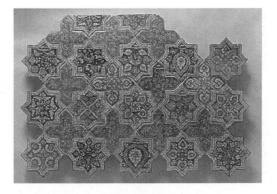

Figure 8.5

A range of other patterns can be made by considering angles other than 45° and continuing the lines into the corner squares. The 60° example is shown below and then again with the corners filled in.

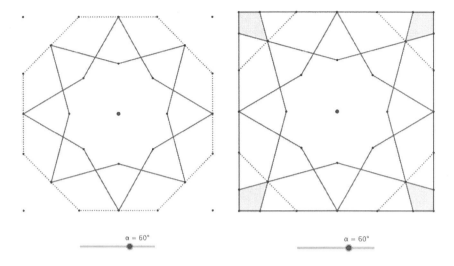

Figure 8.6

This basic square unit can then be tessellated.

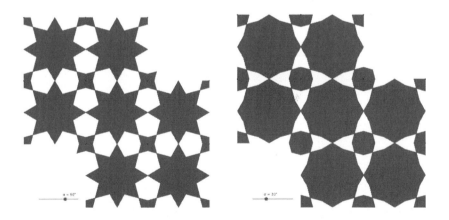

Figure 8.7

We now consider a different underlying grid based on octagons. In this case, our goal is to place eight octagons in a ring around each other as such:

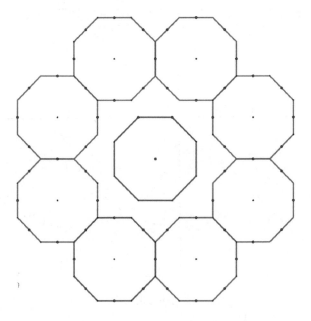

Figure 8.8

While our focus throughout this book rests on using polygonal grids to create overlay patterns, the construction of the grid requires attention as well. The ring of octagons can be constructed within a larger circle as follows:

Begin by constructing a circle and marking off sixteen equal divisions. This can be done by first constructing a horizontal and vertical line through the center of a circle. This creates four right angles at the center. Each of these angles can be bisected to create eight sections and then bisected again to produce sixteen.

Next, create equal-radius circles around the below-labeled points *B* and *D* that intersect at *F*. This line bisects ∠*DAB* into ∠*DAF* and ∠*FAB*.

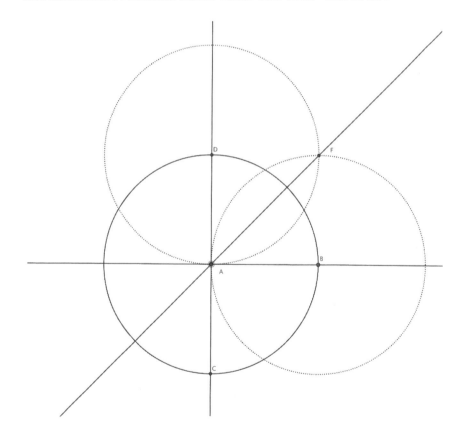

Figure 8.9

Now, identify point *G*, where the ray $\overrightarrow{AF}$ intersects the initial circle *A*. About points *G* and *B*, draw two equal-radii circles that intersect at *I*. Connecting *A* and *I* identifies a new point, *M*. The arc *BM* now represents one-sixteenth of the circle.

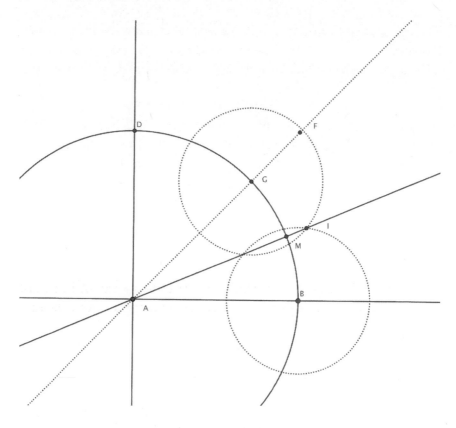

Figure 8.10

There are then several line segments to draw, connecting the various points as shown below.

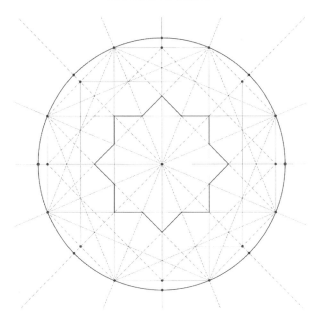

Figure 8.11

Add four pairs of lines, each of which is parallel to the diameter of the original large circle. Their spacing is identified by the large blue polygon in the center.

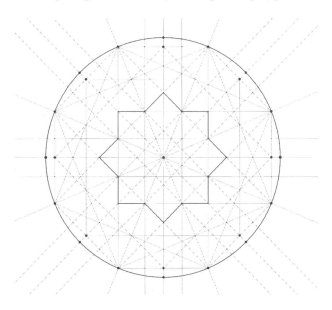

Figure 8.12

Finally, the ring of octagons can be drawn, using the corners that have been identified through all the intersecting lines above.

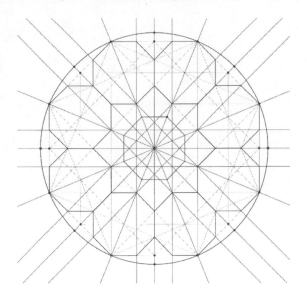

Figure 8.13

Removing the lines leaves just the octagons as shown below.

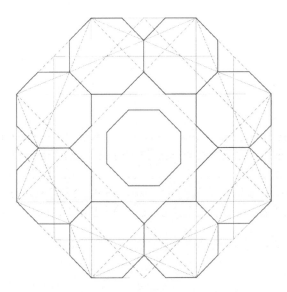

Figure 8.14

The ring of octagons can then be tessellated to fill a region and create an underlying grid, as illustrated below.

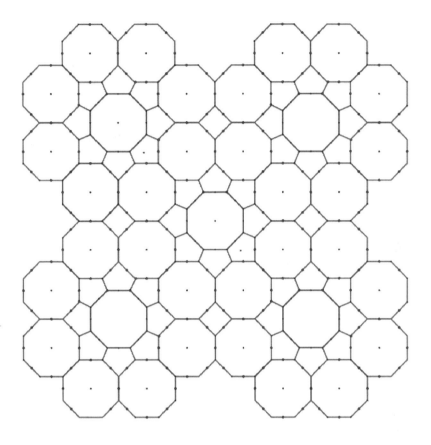

Figure 8.15

Here, the central octagons are connected at the corners to the other octagons, forming five-sided shapes. The centers of all the line segments are identified as well. We are ready to add an overlay tiling. The traditional 45 example is shown below, with a more curvy-looking 30° example. The tile on the top left still shows the underlying octagonal grid as well.

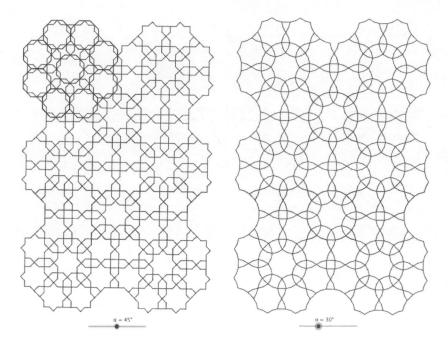

Figure 8.16

Chapter 11
Pentagon Grid

The following rectangle forms the fundamental unit for the tiling we now describe.

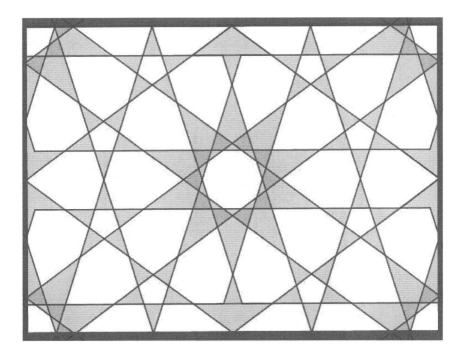

Figure 9.1

The following photo is from the 16th-century shrine in Fatehpur Sikri, India.

Figure 9.2

First, we will create the underlying PIC grid, which, in this case, consists of decagons and pentagons. There will be a central decagon with four pentagons above and below. To the right and left of the central decagon are truncated pentagons. While ten pentagons *can* fit around the decagon, they do not allow for a tiling of the plane; they conflict with a similar grid placed to the left or right. Hence, this grid uses eight pentagons and two truncated pentagons around the central decagon. In each of the four corners of this central unit lies one quarter of another decagon that connects to other tiles to form a whole decagon and, in turn, the center of additional rectangular grids.

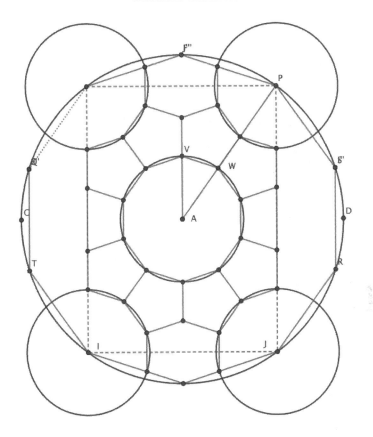

Figure 9.3

We will start this construction by focusing on the outer and inner circles, both centered on point A. The ratio of the radius of the inner circle ($\overline{AV}$) to that of the outer circle ($\overline{AP}$) must be determined such that the pentagons can fit in between the decagons. If we set the inner radius to 1.0, then the length of the side of the central, inner decagon is $1/\varphi$ where $\varphi = (1+\sqrt{5})/2$, often called the golden ratio. Equivalently, $1/\varphi = (\sqrt{5}-1)/2$.

This value is equivalent to the distance between the decagons because it is also a side of a pentagon. Hence, $\overline{AP} = \overline{AV} + \overline{AV}/\varphi + \overline{AV} = \overline{AV} \times (2 + 1/\varphi) = \overline{AV} \times (\sqrt{5}+3)/2$.

In other words, the ratio of the outer radius and inner one is $(\sqrt{5}+3)/2$ to 1.

Now, we are ready to construct this underlying grid. The construction begins with the larger, outer circle (*A*) and an inscribed decagon. To construct an inscribed pentagon, add two smaller circles about points *E* and *G*. Circle *G* is drawn first centered at the midpoint of the below horizontal segment $\overline{AD}$ with radius equal to this mid-length. Circle *E* is then drawn with a radius to be tangent to circle *G* at *H*. The construction is repeated below:

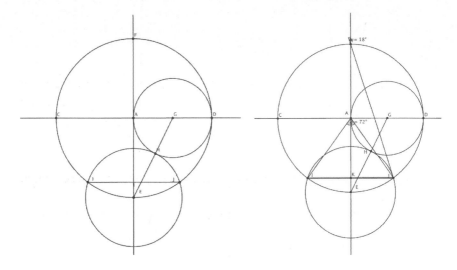

Figure 9.4   Figure 9.5: The complete construction.

Once $\overline{IJ}$ has been identified, it can be copied around the circumference to obtain the pentagon from which we will construct a decagon.

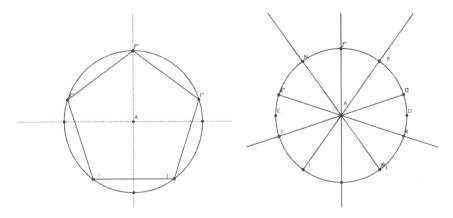

Figure 9.6 and 9.7

To construct the decagon, connect each pentagon vertex with the pentagon's center, and then extend each of these segments through to the other side of the circle. For example, extend $\overline{IA}$ to identify point P and, similarly, with each of the other four points.

Now, connect the ten points on the circumference to reveal the decagon.

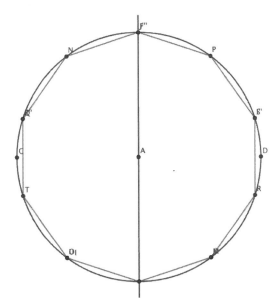

Figure 9.8

Now, we will construct circles around each of the five points N, P, A, I, and J, which will ultimately contain the five smaller decagons of our underlying grid. To determine the correct radius of these circles, we need some auxiliary lines.

Once we have constructed this outer decagon, we need to determine the radius of the outer and inner circles that will hold the smaller decagons. Construct $\overline{NR}$ and $\overline{TP}$. They intersect at point V, which is on the central vertical line. The distance $\overline{AV}$ will be the radius of the five circles.

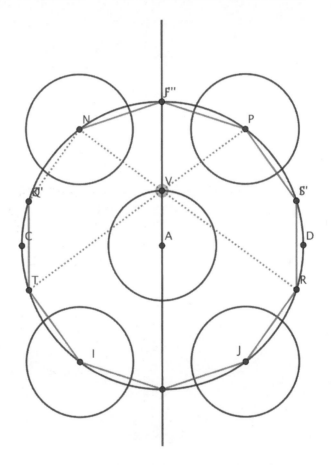

Figure 9.9

At this point, it is helpful to draw in line segments connecting some of the vertices of the large decagon. These lines begin to reveal the pentagons that can be drawn around the inner decagon, as well as the inner decagon itself. For example, OP defines points C_2 and W_1. $\overline{PI}$ define points C_1 and Z. Along with V, found earlier, we now have one of the central pentagons.

It is important to note that the fact that these lines create a pentagon depends on our choice of V. In other words, it is because of the location of V that the distances between the small inner circle and the small outer circles, such as C_1Z, are the same length as the sides of the smaller decagons, such as VC_1.

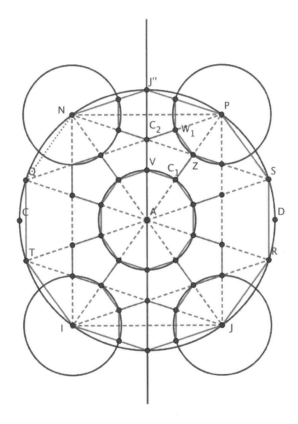

Figure 9.10

With the auxiliary lines removed, we obtain the underlying PIC. To begin the overlay tiling, mark the midpoint of each segment of each pentagon and decagon. Below, the corner decagons have been completed as well, which will make further steps clearer.

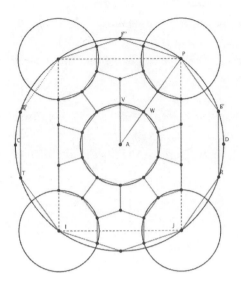

Figure 9.11

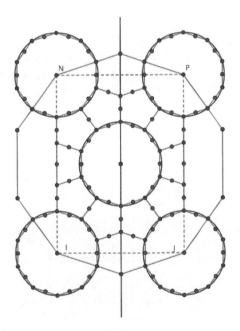

Figure 9.12

Now, draw lines through each of the midpoints as shown. This particular pattern requires those lines to have a **72°** angle.

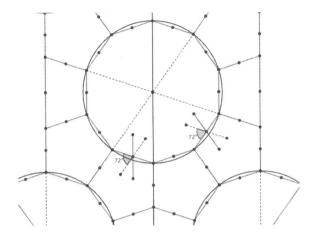

Figure 9.13

As you experiment with this pattern, you will notice several natural landmarks in the grid for determining these 72° angles. For the decagons, a line drawn at 72° from a segment will be parallel to a diameter of the decagon and will terminate exactly at another segment midpoint. (This is unique to the decagon. Not all angles based inside polygons will close up so neatly. But there are plenty of examples to discover.) Reflecting off that midpoint returns the line to yet another midpoint, parallel to a diameter. This continues five times until the line returns to its original position creating a pentagram inside the decagon.

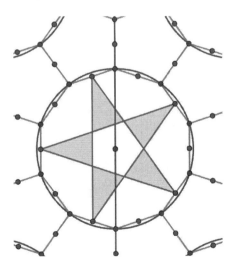

Figure 9.14

Beginning at a midpoint of the decagon creates another overlapping pentagram as well. When the line segments of these pentagrams are extended past the inner decagon, they form parts of the pentagrams that fit inside the other pentagons as illustrated.

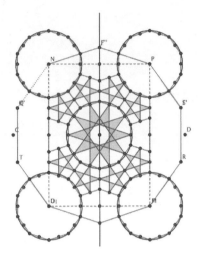

Figure 9.15

The drawing below shows more detail on forming one of the outer pentagrams using a 72° angle. In this drawing, *ABCDE* form an initial pentagon. The midpoints are identified as *F*, *G*, *H*, *I*, and *J*. The rays starting at *J* and *F* intersect at point *P*. Five total intersection points are identified: *L*, *M*, *N*, *O*, and *P*. They connect with the original midpoints to form the perimeter of the pentagram.

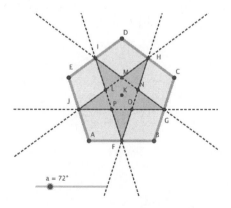

Figure 9.16: A 72° angle.

Alternatively, different angles give rise to different overlay patterns. Decreasing the angle changes the shape of the pentagram by either moving it toward a pentagon or a more acutely-shaped star.

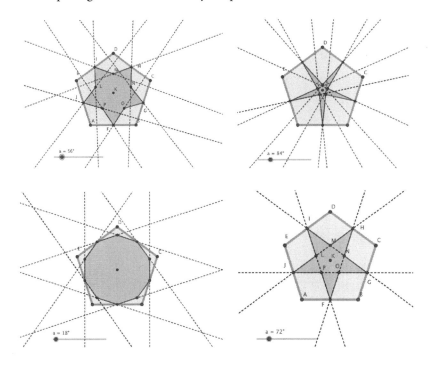

Figure 9.17

The remaining, oddly shaped tiles—on the top, bottom, and the two sides—are similarly treated. Truncate their overlaying line segments to create a rectangular tile, which can be tessellated.

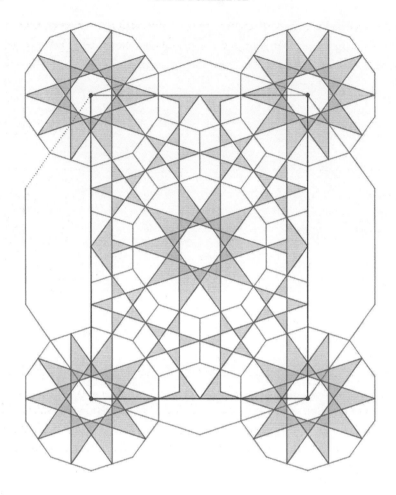

Figure 9.18

The figure below illustrates this PIC grid (center) alongside three different overlays.[2] The "Acute" grid is based on the 72° overlay discussed above. The "middle" uses approximately 50°, and the "obtuse" uses 36°. The two-point design seperates the incident and reflecting rays to create a different type of design. Here, they are based on 72°.

---

[2] Bonner, J., "Three Traditions of Self-Similarity in Fourteenth and Fifteenth Islamic Geometric Ornament."

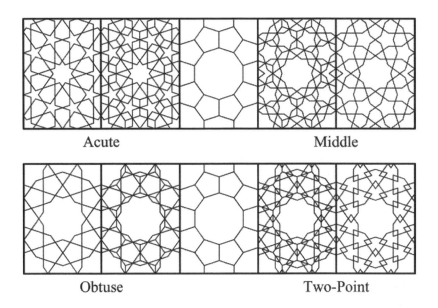

Figure 9.19

All the above overlay angles can easily be determined from points already identified in our underlying grid. As stated above, the 72° version connects a midpoint with an opposing midpoint. A 36° angle can be constructed by connecting midpoints to the closest midpoint on either side. A 50° angle is found by connecting a midpoint with an opposing midpoint as illustrated below.

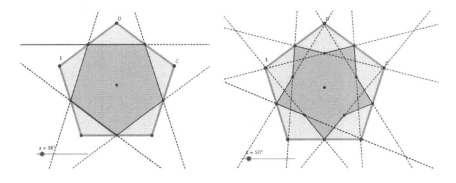

Figure 9.20

A completed pattern can be found in the appendix, along with a midpoint grid from which to explore other angles.

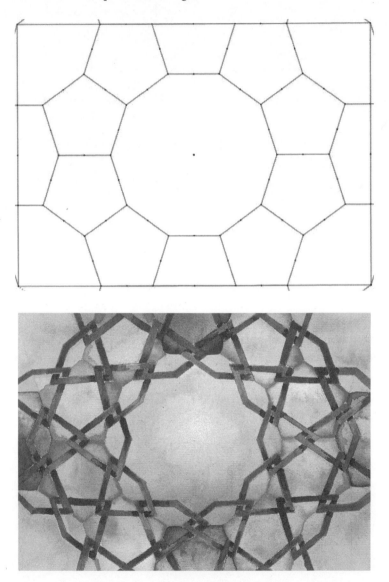

We can create a different pentagonal design by placing five decagons around a central decagon as follows: Similar to the design above, begin with a decagon, and use the two dotted lines to identify the inner radius for another smaller decagon.

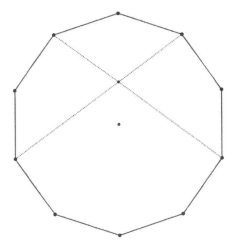

Figure 9.21

About the center, construct a circle of that radius and proceed to inscribe a smaller decagon. The dotted diagonals of the larger decagon establish the vertices of this smaller decagon. Then construct a similar sized decagon at the five corners shown.

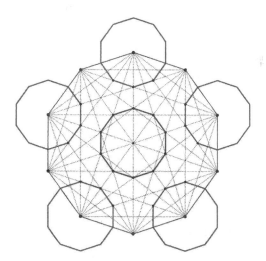

Figure 9.22

Some additional points are indicated that begin to make the pentagons around the central decagon visible.

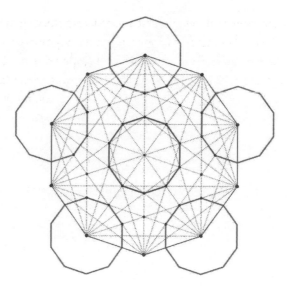

Figure 9.23

The final grid is shown below along with a pentagonal border. The midpoints of the segments are also marked.

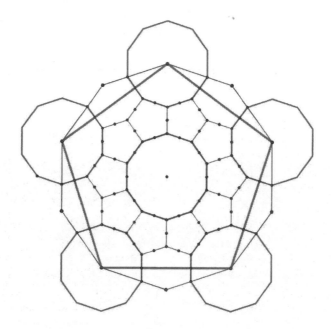

Figure 9.24

A pattern can now be drawn by first constructing the ring of five-pointed stars around the central decagon. By connecting the midpoints of the pentagons as illustrated, these pentagrams are formed. The lines making up these stars can be extended to form the central ten-pointed star in the middle. Extending these lines outward begins to illustrate how the pieces on the boundary get filled in.

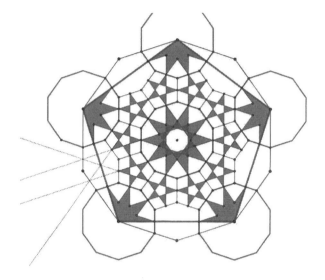

Figure 9.25

The final pattern, based on 72°, appears now with the underlying grid as well.

Figure 9.26

An alternative pattern can be constructed using a two-point overlay. These types of patterns employ the same rules as above, but incidence angles are drawn through *two* points on each segment, identified by dividing it in thirds. A final two-point overlay is shown below in black, with the original polygon grid indicated in brown.

Figure 9.27

To construct the two-point grid, return to your underlying pattern and divide each side in thirds to identify two points, shown in blue. Only one of the points is shown which will be used to construct the inner pattern.

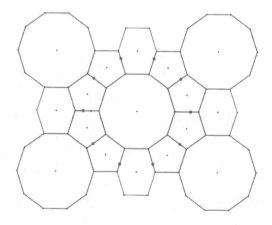

Figure 9.28

From the blue points identified above, construct a ten-pointed star.

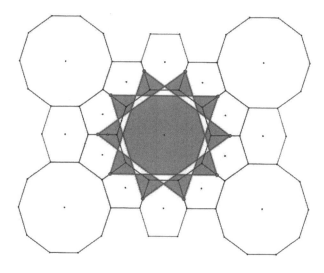

Figure 9.29

Next, construct a ten-pointed rosette (a typical flower-shaped pattern) inside the shaded dodecagon, which was created by the ten-pointed star. Below, the dotted lines indicate where to locate the vertices of the petal. Note that *B7* and *A7* are one-third points along the hexagon sides. Connecting them and noticing their intersection with the ten-pointed star identifies a pair of points on each side, example *V6*.

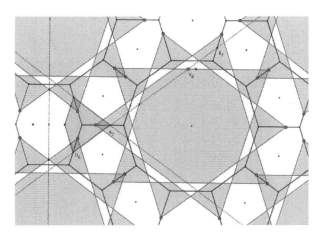

Figure 9.30

Connect these points (e.g., *V6*) to create five pairs of parallel lines, which identify the petals of the rosette.

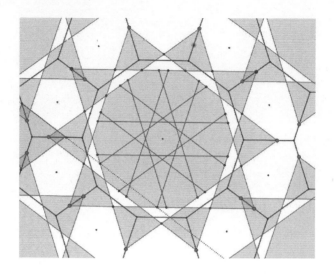

Figure 9.31

Below, I highlighted the rosette and the outer ten-pointed star, as well as an additional quadrilateral on the left, which will be needed. Note that it too is constructed by identifying and connecting one-third points on the other polygonal tiles.

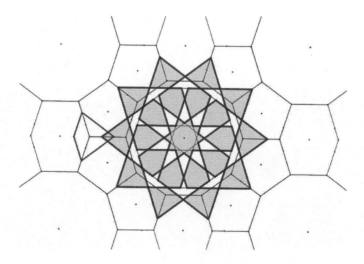

Figure 9.32

The remainder of the picture is obtained by repeating this construction around the other dodecagonal tiles.

Figure 9.33

An example of this pattern is on display at the Metropolitan Museum of Art in New York.

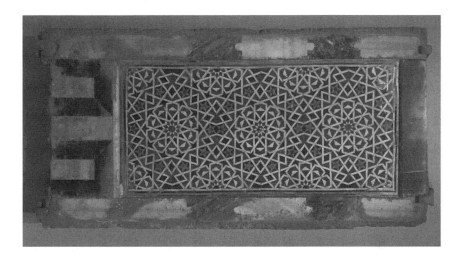

Figure 9.34

Chapter 12

Dodecagon-Triangle-Square Grid

An early historical example was illustrated by Hankin that incorporates dodecagons, triangles, squares, and other shapes to fill in a rectangular region that can thus be tessellated.[3]

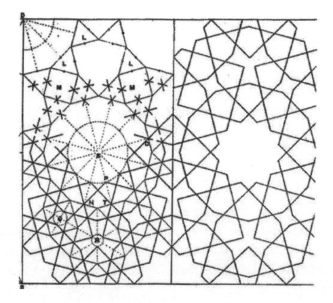

Figure 10.1

The construction begins with a circle divided into twelve points that form the vertices of a dodecagon.

---

[3] Hankin, E. H., "The Drawing of Geometric Patterns in Sacred Art," Memoirs of the Archeological Survey of India, 1925.

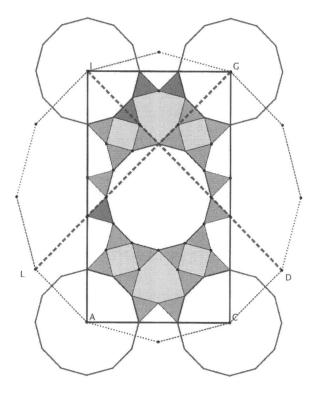

Figure 10.2

Once the twelve points are identified, the tessellating rectangle is given by the points *ACGI*. Much like the two previous constructions, we next identify the size of the five smaller dodecagons that are in the center and four corners. Their radius can be found by considering the two diagonal green lines connecting points on the outer dodecagon, $\overline{ID}$ and $\overline{GL}$. Their intersection defines the inner dodecagon's radius. Use this same radius to draw all five dodecagons.

Now, construct equilateral triangles off each side of each dodecagon. These are illustrated by the purples shapes below. The remaining spaces, the squares and oddly shaped hexagons, stay as they are.

The final grid is shown below.

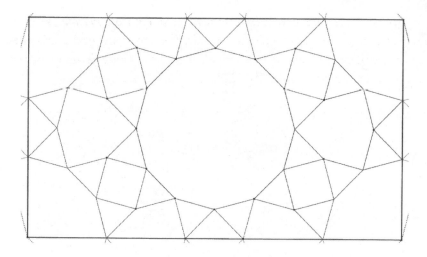

Proceeding as usual, draw line segments through the midpoints of each side to produce the final pattern. Below are some examples for given angles.

Figure 10.3: 15° overlays, from left to right.

For good measure, here is the original illustration from Hankin, showing the 60° pattern. The middle picture includes the underlying grid of triangles, squares, dodecagons, and oddly shaped hexagons. The right-most picture shows the final design with the grid removed.

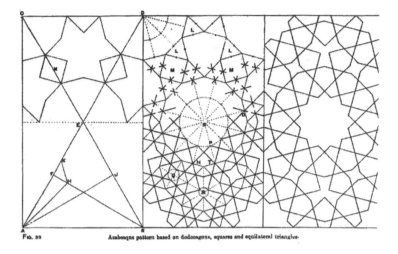

Figure 10.4

Alternatively, the patterns can be drawn with just the outline of the polygons leaving the interiors blank.

Rather than constructing a pattern within a rectangle, the outer dodecagon can be used to create a square and hexagon based pattern as follows:

Once the dodecagon is drawn, the secondary radius is found through the following construction. This produces a different radius from which four corner circles are found.

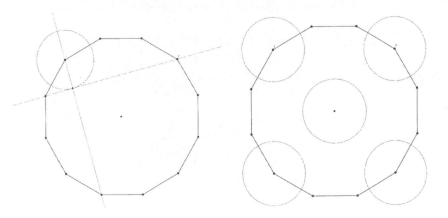

Figure 10.5   Figure 10.6

Proceeding similarly, dodecagons are inscribed within these circles, upon which equilateral triangles are built. The remaining squares and hexagons appear from these edges.

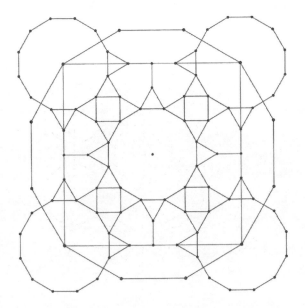

Figure 10.7

Lines drawn through the midpoints at 45-degrees create the pattern below.

Figure 10.8

Another grid can be constructed using six small outer dodecagons, but the spacing is different. The smaller radii are identified by drawing a different pair of lines through the dodecagon. Note that one of the lines is a diameter; the other is not.

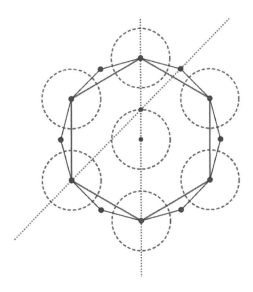

Figure 10.9

The intersection of these lines defines the smaller radius. Draw a circle of this radius at the center and about six equally-spaced points on the perimeter. Construct dodecagons in each circle. The lines shown, which connect points of the outer dodecagon, identify the corners of these smaller dodecagons.

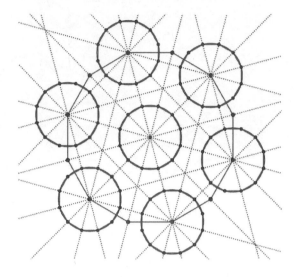

Figure 10.7

From there, construct equilateral triangles on the edges of the dodecagons. The squares and shield shapes automatically appear as shown. The individual tile, which we can tessellate to create a grid, is hexagonal.

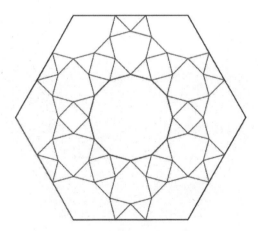

Figure 10.8

Find the midpoints, select an incidence angle, and overlay.

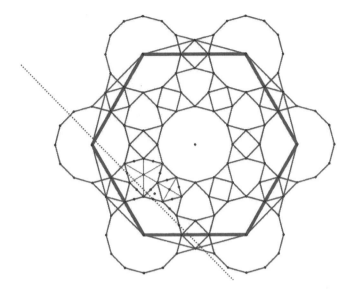

Figure 10.9

A complete pattern appears below with 60° incidence angles. The original grid appears in brown and the overlay in blue.

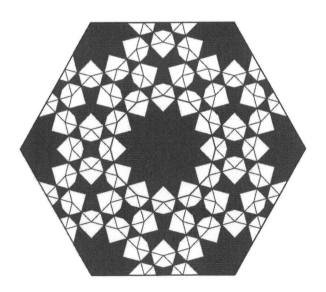

Figure 10.10

Here we just see the outlines the overlay tiles.

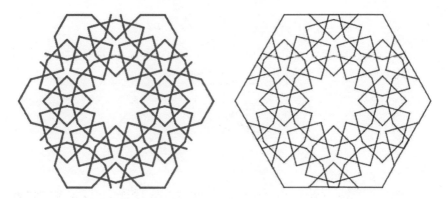

Figure 10.11

Chapter 13

3-3-4-3-4 Grid

The following grid merges the 3-3-4-3-4 grid with the 6-4-3 grid to create the following polygonal grid:

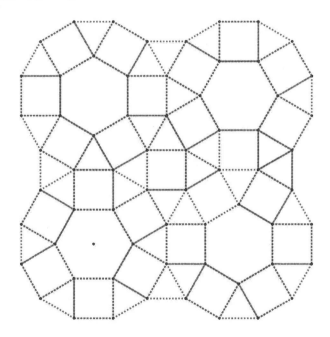

Figure 11.1

Proceed with incidence lines along all the edges, we get the following results: Below, you can see the development of 30°, 45°, 60°, and 75° overlay grids. First, the incidence overlay on top of a base grid, then a shaded overlay grid with the base grid still visible, and finally, the overlay tiling without the underlying grid.

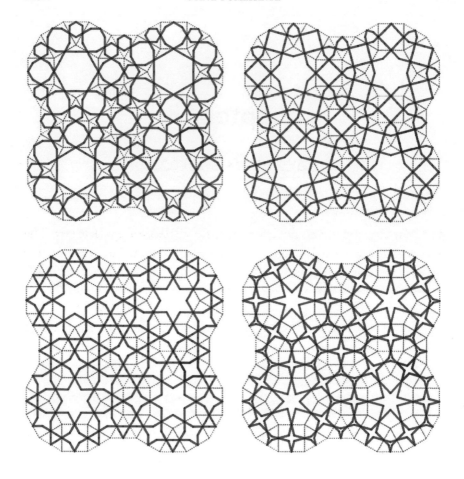

Chapter 14

Dodecagon-Pentagon Grid

To create the underlying grid, first divide a circle into 24 segments. The repeating unit will be the rectangle shown below. The radius of the smaller circles is identified by the intersection of the 45° line through the center with the vertical side of the rectangle as illustrated by the solid green line.

Once this radius is determined, a dodecagon can be constructed about each corner as well as the center.

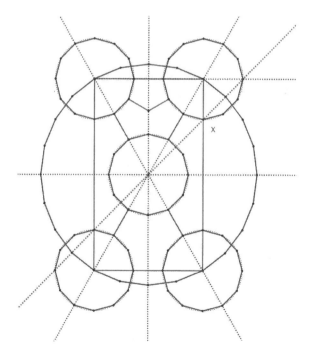

Figure 12.1

The grid is completed now by extending rays through the centers of these circles to create a set of pentagons in between the dodecagons.

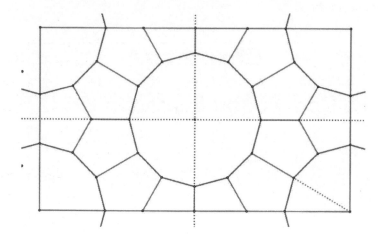

Figure 12.2

Now, the overlay design is constructed by choosing an incidence angle. An angle of 75° will produce parallel rosettes as the construction begins below. As shown, there is a pair of incidence and reflected lines crossing each edge of the central dodecagon. These are used to define the central rosette. This pattern can then be translated to the four corners of the rectangle to extend the pattern.

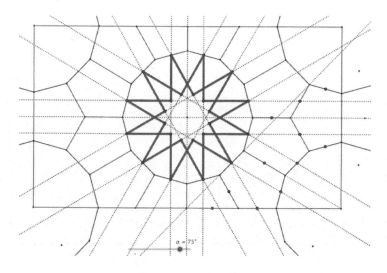

Figure 12.3: A 75° overlay

Next, the tiles inside the pentagons can be constructed. Some of the incidence lines have already been defined by the dodecagon steps above; others have to be drawn.

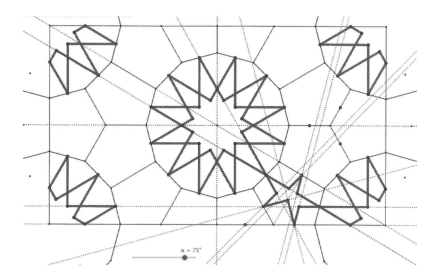

Figure 12.4: A 75° overlay

The remaining pentagons can be filled in by repeated application of these incidence-reflection lines or by suitable reflections of the tile illustrated above.

Figure 12.5

Note that because the pentagons have unequal sides, the obtained five-pointed stars are a little odd. We can improve this a little by breaking one of our rules. We choose slightly different "midpoints" to continue the construction. Using a circle about $L1$ to measure the midpoint of the segment connecting the dodecagons, mark off similar distance from $R1$, $V3$, and $C3$ to obtain "midpoints" for the three bolder segments.

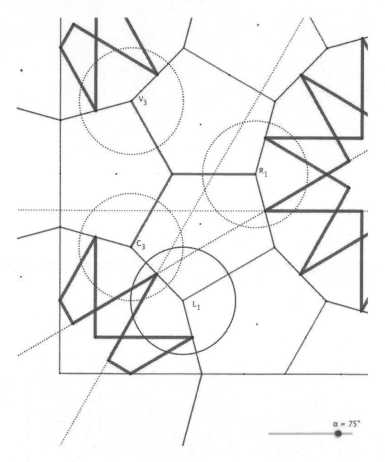

Figure 12.6: Continuing with the 75° overlay.

Then use these new points to establish incident and reflection lines and the resultant five-pointed star. Note that four of the five edges of this star form boundaries of the rosette petals which are now all of equal length.

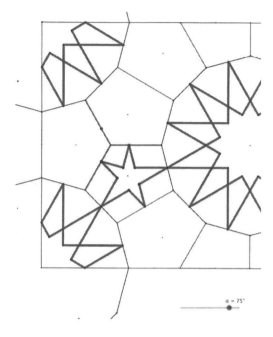

Figure 12.7

By additional constructions or reflections, these five-pointed stars can be replicated throughout the pattern.

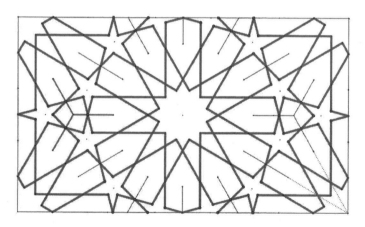

Figure 12.8

And finally, without the underlying grid and repeated with parallel, convergent and divergent rosettes:

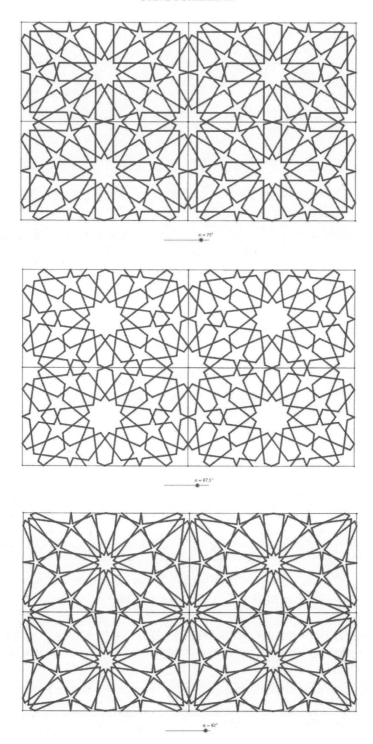

Chapter 15

The Rosette

The rosette is one of the most popular architectural designs in Islamic art. Although this shape can be constructed using the PIC method, here, we will approach it differently in an effort to emphasize specific characteristics of the design. The study of rosettes, from a mathematical perspective, provides for an endless study of ratios, proportion, bisectors, angles, symmetry, and parallelism. In this chapter, we will construct a rosette with proportions that emphasize those elements.

By varying the orientation of the sides of a rosette, we can create three distinct but related shapes: divergent, parallel, and convergent.

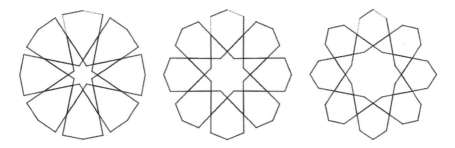

Figure 13.1: The three rosette shapes.

An important aesthetic component of the rosette is the equivalent lengths of the upper segment and side segment (as illustrated by the dotted lines), which provides an important hint on their construction. This is not a PIC pattern but is a classical motif and relies upon many geometric principles.

We begin by constructing an eight-petalled, parallel rosette, the simplest case. First, construct a circle and divide it into sixteenths.

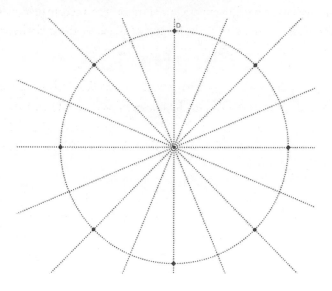

Figure 13.2

Connect the eight points shown above to form an octagon, and then construct tangent lines at each of the points.

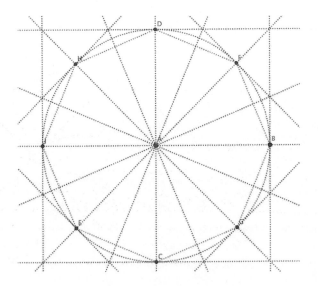

Figure 13.3

Identify the intersection points of the tangent lines and construct a circle about each intersection point. Each circle must be tangent to the radial lines that go through the two closest octagon vertices. The picture below illustrates four such circles, centered at K, R, O, and N. Each of these circles intersects with a diameter of the original circle at S, U, B, or Z.

Next, join points U and Z, and extend in this line in both directions so that it intersects with two sides of the octagon, $\overline{DH}$ (at $C1$) and $\overline{EC}$ (at $D1$). In this manner, the line segment $\overline{C1U}$ is the left side of a petal and $\overline{C1D}$ is the top half of a petal. It is not obvious yet that they are of equal length, but we will prove that below.

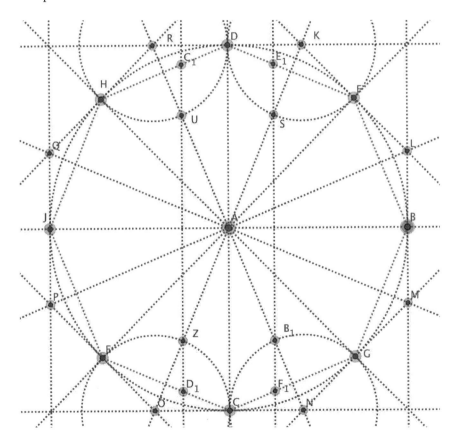

Figure 13.4

Repeat this process three more times to create three more sets of parallel lines. This is done using circles at *Q, P, M, L* then *K, L, O, P* and then *Q, R, M, N*. Each of these sets of parallel lines intersects the octagon at points that are vertices of the petals. The picture below removes the diameters and tangent lines to better illustrate these sets of parallel lines.

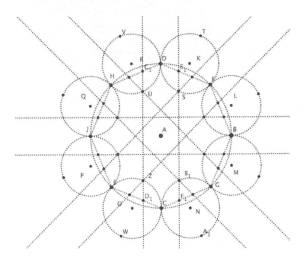

Figure 13.5

All the vertices of the petals are now identified and shown below. We have created a parallel rosette.

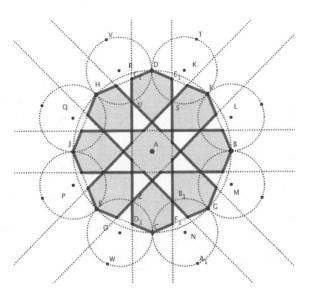

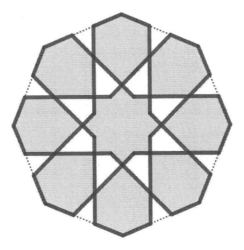

Figure 13.6

The construction of convergent and divergent rosettes requires some additional intermediate steps. These constructions will make clear why the tops and sides are equal in length and how the parallel rosette is just an example of the more general construction.

We will begin with the construction of a convergent rosette. First, we will have to create some additional lines. The main addition is in constructing a bisector of angle ∠DKS as shown. (The dotted red line divides that angle into two equal components.) Note that $\overline{DK}$ and $\overline{KS}$ have equal length, both being radii of the same circle.

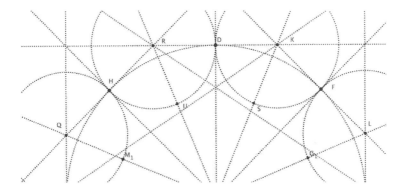

Figure 13.7

Next, we can pick any point on the bisector to continue this construction. To do so, we arbitrarily draw a line through D at an angle to the tangent. If we choose a 22.5° angle, we will obtain the parallel rosette above, so we will use a different value, here of approximately 30°. The intersection of this line with the bisector (denoted by C_1 and D_1) form the upper corners of the petal. Line segments $\overline{C_1S}$ and $\overline{D_1U}$ are its sides. Their equivalent length follows from the side-angle-side congruence principle. $\overline{DK}$ and $\overline{KS}$ are equal because they are radii of the same circle. $\overline{KC_1}$ is equal to itself, and angle $\angle DKC_1$ is equal to $\angle C_1KS$ because they were found by bisecting the angle $\angle DKS$.

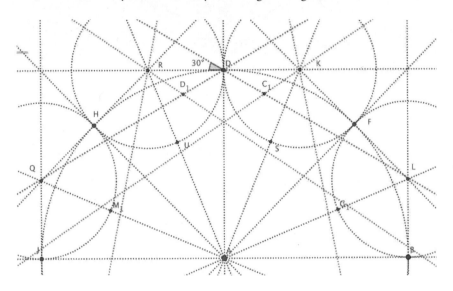

Figure 13.8

All that remains to be identified is the lower point of the petal. To find this, we repeat the steps above about point F by constructing a line at 30° to the tangent at F and by bisecting $\angle FKS$. These two lines intersect at E_1 below. E_1 will be a vertex of the next petal, but it is also needed to locate the bottom of the first petal as shown. Extending $\overline{E_1S}$ to the vertical diameter of the original circle identifies F_1, which allows us to construct one complete petal.

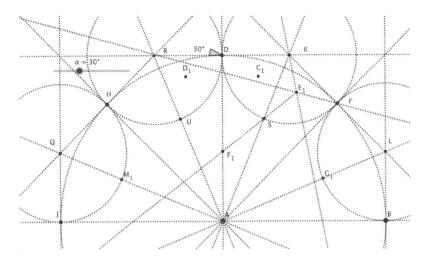

Figure 13.9

Connect $\overline{DC_1SF_1UD_1}$ to form the top petal. Note that the segments $\overline{SF_1}$ and $\overline{UF_1}$ are extended to the next radial lines. This will be helpful in forming the inside of the rosette.

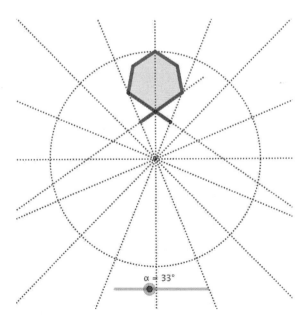

Figure 13.10: A **33°** angle

Repeat this procedure around each of the eight vertices of the original octagon to achieve the following design. Note that these convergent rosettes are formed from an incident line obtained by connecting intersections of the original circle with various radii, thus not requiring the use of a protractor measure out a specific angle.

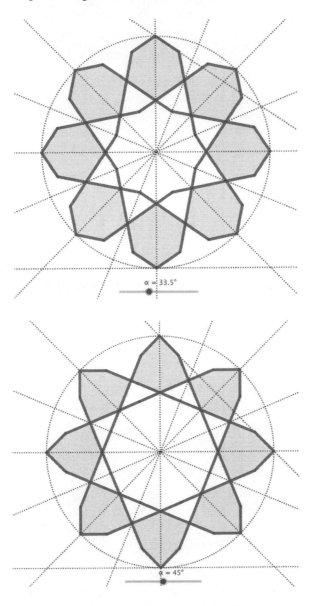

Figure 13.11: **33.5°** and **45°** angles, from left to right.

Finally, divergent rosettes can be formed using similar points.

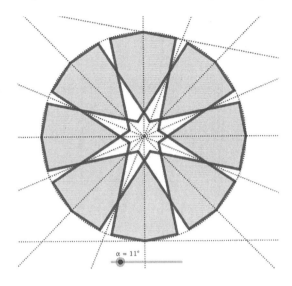

Figure 13.12: A divergent rosette at 11°.

The parallel rosette, which we initially constructed using an intersection point, is shown below as just another choice of angle, namely 22.5°.

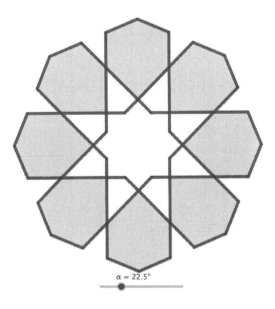

Figure 13.13: A parallel rosette at 22.5°.

It is also possible to construct rosettes with the usual PIC method. The basic idea is to construct the PIC grid simultaneously with the pattern. Begin with an octagon as illustrated below. Bisect a side $\overline{DE}$. Construct an arc centered at a vertex E with radius equal to $\overline{EJ}$, i.e., one half of the length of the side.

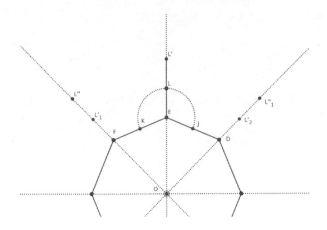

Figure 13.14

This arc defines the point L, which is on a ray through the center of the octagon. Mark that same distance away from L to define L'. Repeat this procedure at each vertex of the original octagon. Connect the newly created points as below. The underlying grid is the ring of quadrilaterals around the original octagon, and the midpoints are already identified.

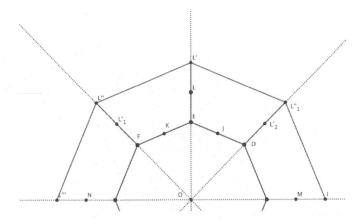

Figure 13.15

The incidence lines can be drawn through *J* and *L* as illustrated below.

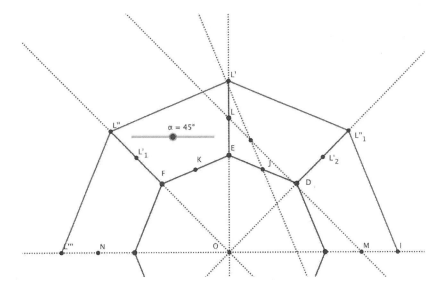

Figure 13.16

Construct the rosette petal using the incidence lines as shown.

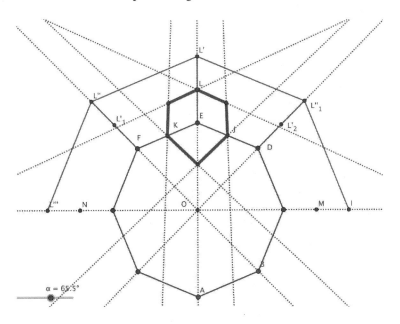

Figure 13.17

And repeated at each quadrilateral to reveal the rosette.

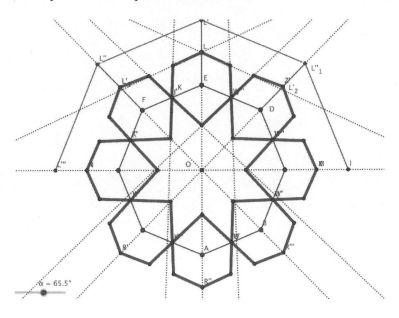

Figure 13.18

Illustrated below is the final rosette with a piece of the PIC grid underneath and a view without the grid.

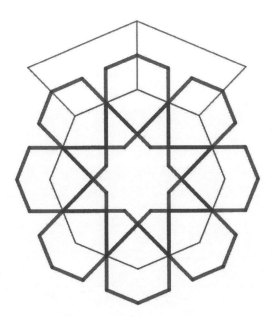

Classical Geometry

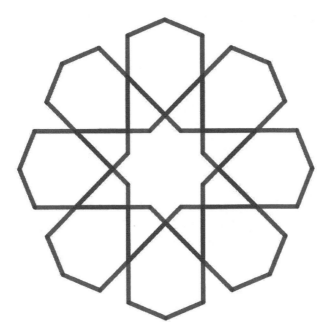

Figure 13.19

Chapter 16

Dodecagon-Hexagon-Square Grid

In this chapter, we overlay patterns on a grid of dodecagons, hexagons, and squares. To begin, we will first construct the 12-3-6 grid. Next, we will overlay patterns using the PIC method.

Begin with a horizontal line and three circles as shown below. This defines the six points on the circumference of the center circle.

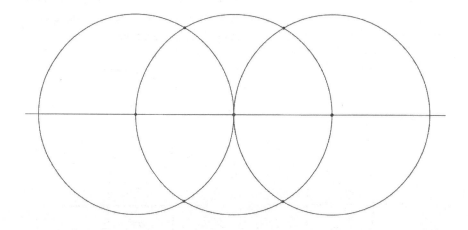

Figure 14.1

Now, several lines are drawn for reference. First, insert diagonal lines, which connect points on the center circle, then proceed to draw the inner segments making a six-pointed star. Below, the central rectangle is bolded—it will contain the final tiling unit.

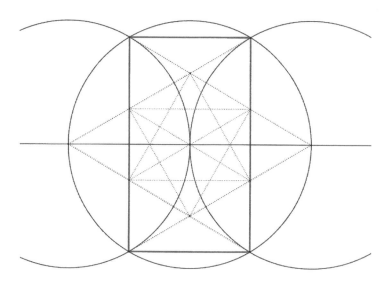

Figure 14.2

The size of the hexagon can now be found as follows: First, draw two circles about the midpoints of the two vertical edges of the unit. Then draw a circle, centered in the middle, that is tangent to these circles.

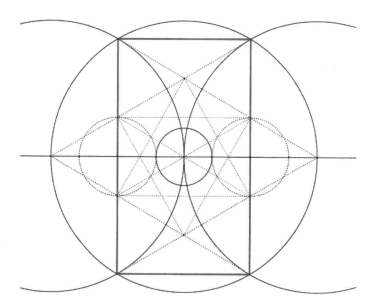

Figure 14.3

Draw circles of that same radius at each of the six points shown below.

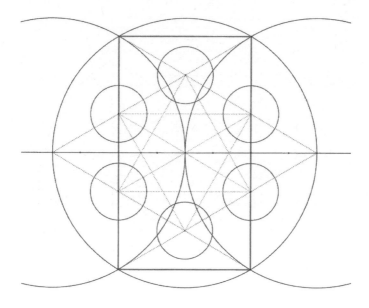

Figure 14.4

Using the intersection points between these circles and the previously drawn reference lines, we can identify the sides of hexagons to inscribe.

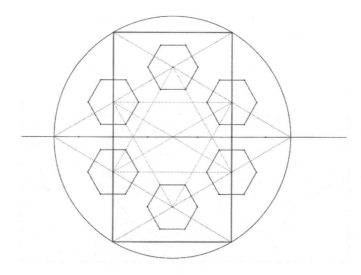

Figure 14.5

Now, fill in the squares between the hexagons, and the dodecagons begin to appear.

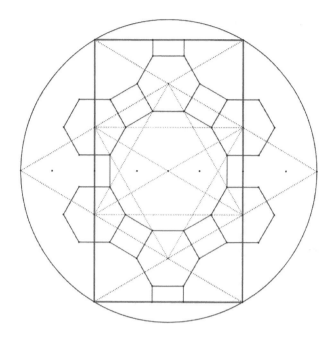

Figure 14.6

The final underlying tiling unit appears below. The reference lines have been removed and the midpoints marked.

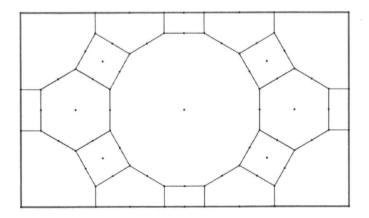

Figure 14.7

Several incidence angles fit the underlying pattern particularly well. For example, 63.5° connects the midpoint of a square side with an opposing corner.

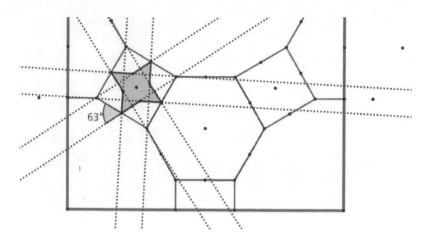

Figure 14.8

It is helpful to notice that several vertices and midpoints correspond to specific angles for each shape.

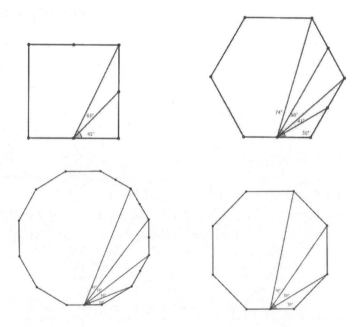

Figure 14.9

The final overlay grid can be seen below with the underlying grid still visible.

Chapter 17
Topkapi Scroll

One of the oldest and most famous patterns is preserved in the Topkapi Scrolls, dating back to the fifteenth century. An example below illustrates an underlying grid in red and the overlay pattern in black. It is one of the earliest (and few) historical pieces that evidence this method of construction. While not based on a regular tiling, it clearly illustrates a grid of basic shapes, pentagons, dodecagons, and trapezoids that are replaced with the overlying pattern.

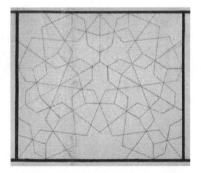

Figure 15.1

This grid is comprised of pentagons, dodecagons, and trapezoids, but we begin with a hexagon; the corners and center of which center the dodecagons. It is interesting to note that what appear to be regular hexagons are not and slightly different. Careful measurement of the sides reveals they are not equal. Where the three meet, the interior angle would have to be 120°, but regular hexagons have an interior angle of 108°. Nevertheless, the pattern is esthetic and of historical importance. The topkapi scrolls are some of the historical evidence for the polygon method.

The inner radius is found through a combination of several steps. The first is to construct the two diagonals shown and observe their intersection. Then bisect the corner angle and observe the intersection of the bisector with the vertical as shown.

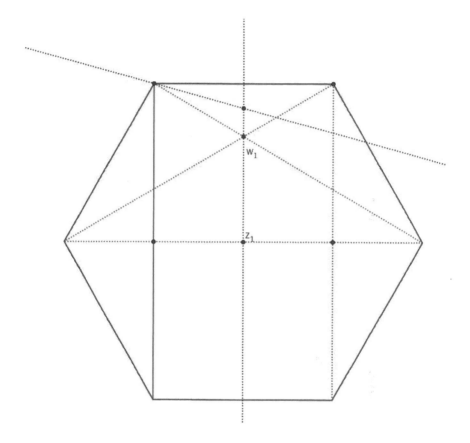

Figure 15.2

Now, draw three circles, the first around W_1, then two others of equal radius about the points I and L, where the first circle intersected the diagonals. Then connect I and L with the center of the hexagon and observe the intersection of these segments with the two circles. These two points lie on the circumference of the center circle. The usage of these circle is to force some of the edges of the pentagons to be equal in length.

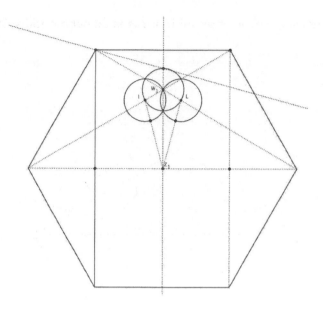

Figure 15.3

The center dodecagon is inscribed within this circle. Dodecagons in the corners intersect the angle bisectors at the remaining points of the hexagons. Some additional line segments begin to realize the underlying grid.

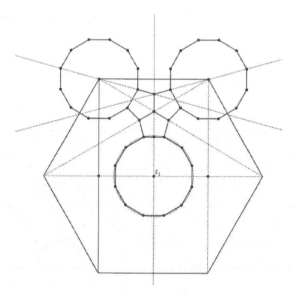

Figure 15.4

Additional diagonals of the initial hexagon fill out the grid as follows:

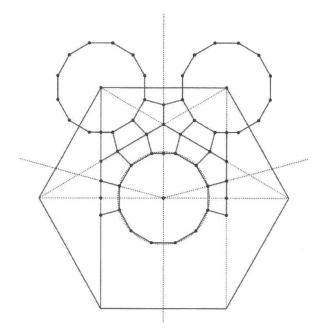

Figure 15.5

The complete grid is shown below with the segment midpoints identified.

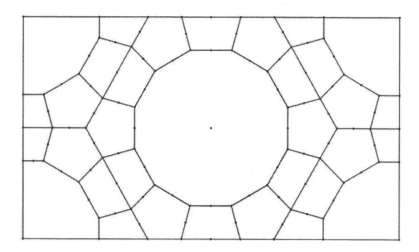

Figure 15.6

The traditional design uses an incidence angle of approximately 65-69° and produces slightly convergent petals on the rosettes.

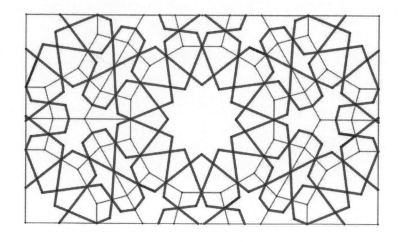

Figure 15.7

Figure 15.8 below displays angles of 54°, which is more convergent, and 75° which produces parallel rosettes.

Figure 15.8

CLASSICAL GEOMETRY 139

A slightly more complicated pattern from the Topkapi is illustrated below. The underlying grid consists of dodecagons, rhombi, and trapezoids with an incidence angle of 60°.

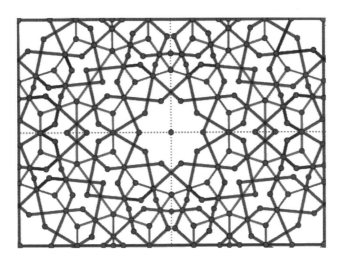

Figure 15.9

Chapter 18
Jali Screen

While strictly not a PIC grid, the following is an example of a jali screen. These constructions simultaneously served as windows and decoration.

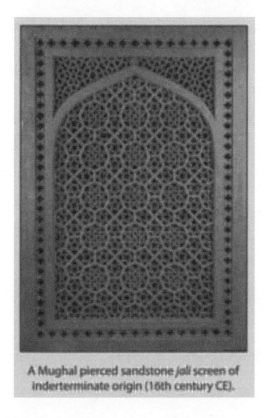

A Mughal pierced sandstone jali screen of indeterminate origin (16th century CE).

Begin with a circle with a horizontal and perpendicular drawn through the center. Bisecting these lines identifies the corners of an octagon inside the circle.

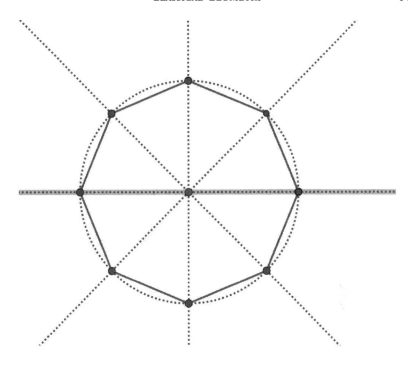

Figure 16.2

To determine the appropriate spacing, construct a circle about the rightmost point of the octagon with a radius equal to the length of a side. This will identify the leftmost point of a new octagon to the right of the original. Next, construct a circle of radius equal to the original circle that is tangent to that point.

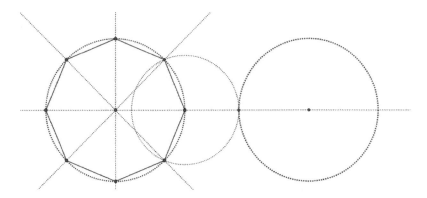

Figure 16.3

Inside the new circle, construct a similar octagon, using a perpendicular and angle bisectors to identify the corners.

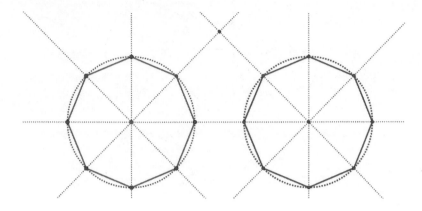

Figure 16.4

The intersection of the diagonal and vertical lines identify the centers of new circles that are drawn above the two just constructed.

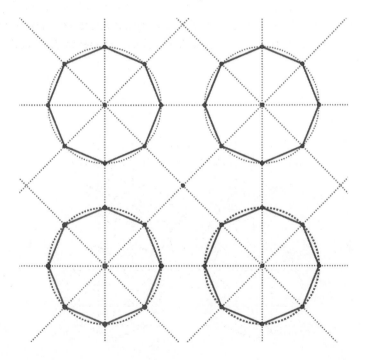

Figure 16.5

The central unit is now completed by adding segments as shown.

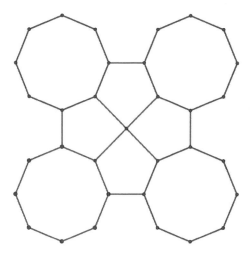

Figure 16.6

Unlike other patterns, the overlay pattern is created without removing this underlying grid as illustrated below. The overlay pattern shown is constructed by connecting the midpoints of the polygonal units to reveal five- and eight-pointed stars.

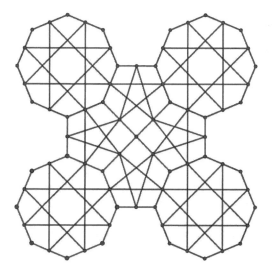

Figure 16.7

The screen shown at the beginning of this section can now be created by tessellating the above pattern.

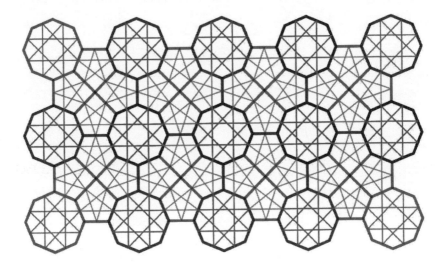

Figure 16.8

Chapter 19
Dodecagon-Triangle Grid

In this chapter, the central unit that fills in the dodecagon will ultimately look like the figure below.

Figure 17.1

To begin, construct a dodecagon with its radii. An inner dodecagon is also drawn. The size of the inner dodecagon is determined by the chord drawn that connects two of the outer points.

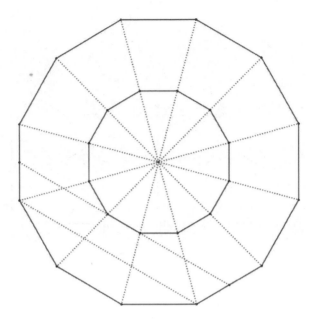

Figure 17.2

On the sides of this inner dodecagon, erect equilateral triangles. Now, construct reflected triangles at the top, bottom, and sides to complete the underlying grid, solely made of triangles and odd-shaped hexagons.

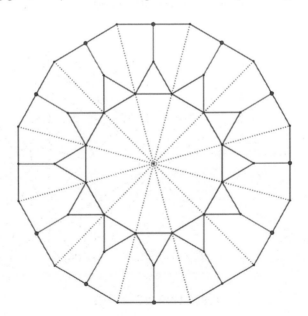

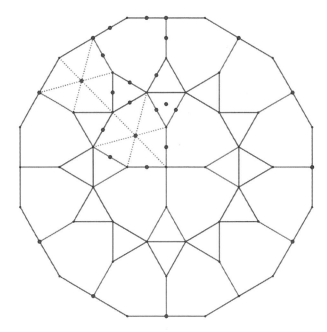

Figure 17.3

The PIC method now constructs shapes in each of these regions. For example, a 30° angle in the triangles is illustrated below: Similarly, the hexagons are filled and replicated around the grid.

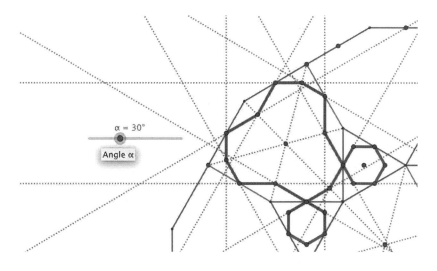

Figure 17.4: A 30° angle

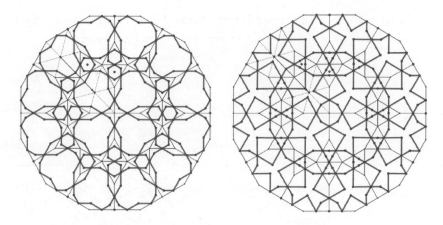

Figure 17.5: The **30°** and **60°** angle versions.

From here, we can replicate the divided dodecagon above by first identifying a triangle that is exterior to the dodecagon on one of its sides and then repeating the constructions above.

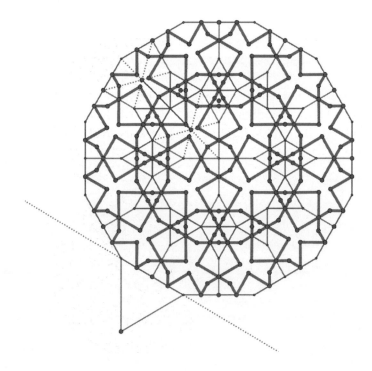

Figure 17.6

What remains is the center triangle. Divide this triangle into four similar equilateral triangles, and now, extend the earlier lines into these:

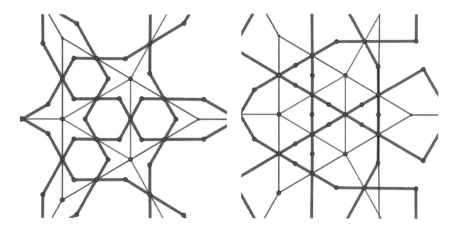

Figure 17.7: A close-up of the center triangle, 30° and 60° angles.

The final 60° tiling follows.

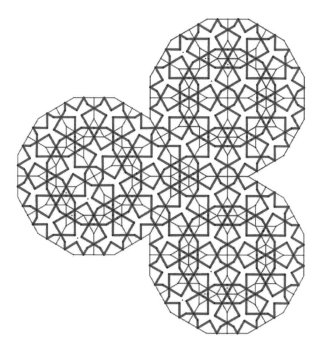

Figure 17.8

Note that the interior dodecagon itself contains four triangles and four hexagons. This shape itself can be tessellated with an additional triangle where three of them meet up as illustrated below:

Figure 17.9

Appendix 1
Even More Tilings

Here is the 3-3-3-4-3 tiling.

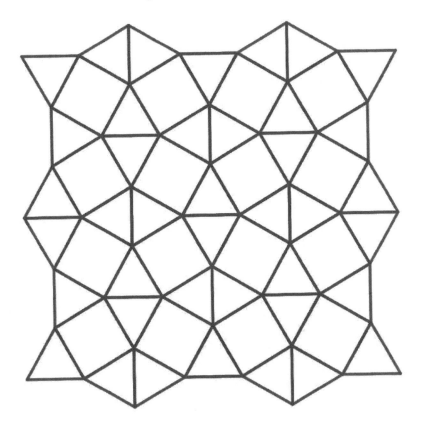

With overlays made from 45° and 60° angles.

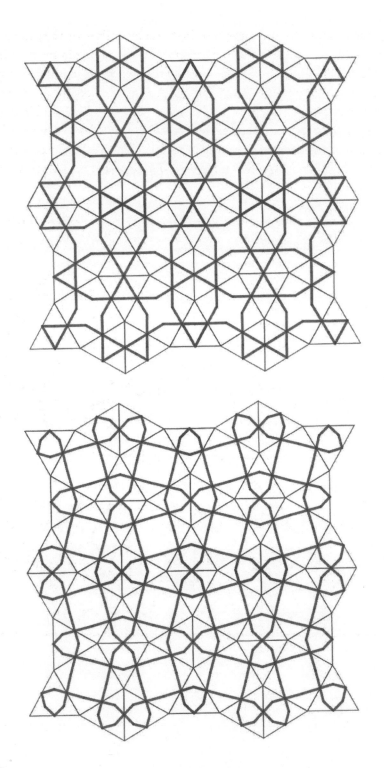

Here is an example that uses an octagon in the center, surrounded by pentagons. Then dodecagons are placed in the corners with filler pieces in the middle of the sides. This is not an exact fit because the two pentagons and octagon only have interior angles of 351°, not 360°. Also, the dodecagon-pentagon vertices would require 366°. As such, they have to be adjusted a bit, which was historically common. The final results, though, are pleasing, nonetheless, but give another direction in which this method can be extended. The two examples shown are, respectively, 56° and 67°. Note that the inner rosette is, respectively, convergent and parallel.

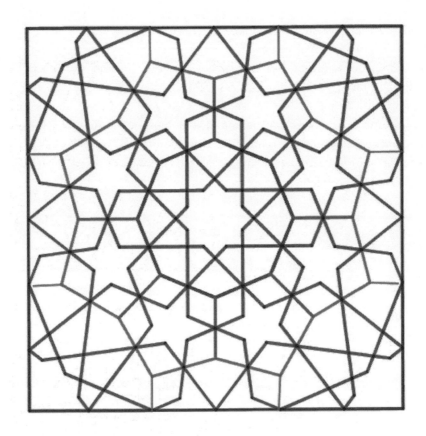

Here is a pattern with a decagon center and corner with assorted trapezoids and fillers.

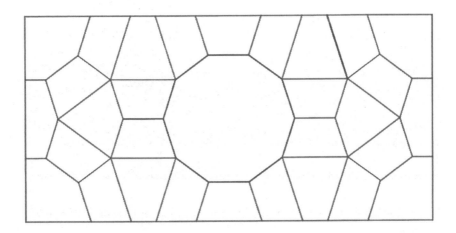

With a 72° overlay, we get the following design:

Here is an alternative still with a center decagon but using trapezoids and rhombi to fill in the tiling rectangle.

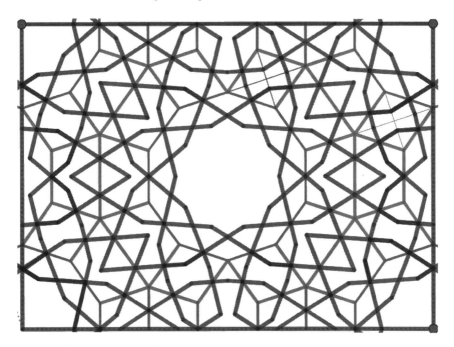

The following patterns make use of hexagons and rhombi using 60° and 72° overlays.

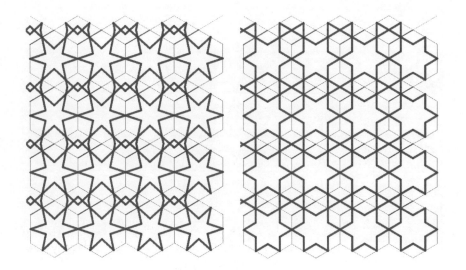

Appendix 2
Completed Underlying Tilings

In the following pages, you will find a number of completed underlying grids. These may be used to skip the initial grid construction and jump directly into practicing the PIC method and secondary tilings.

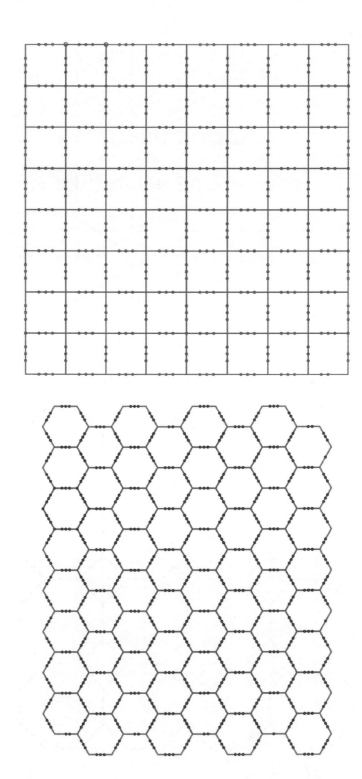

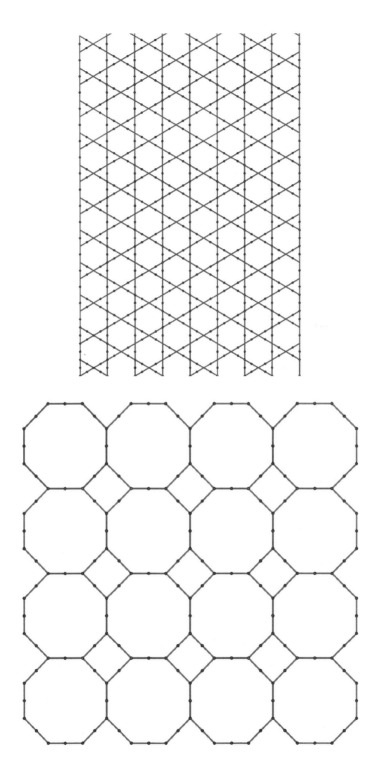

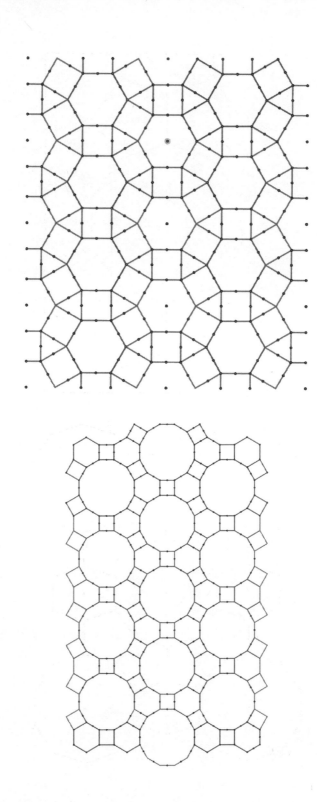

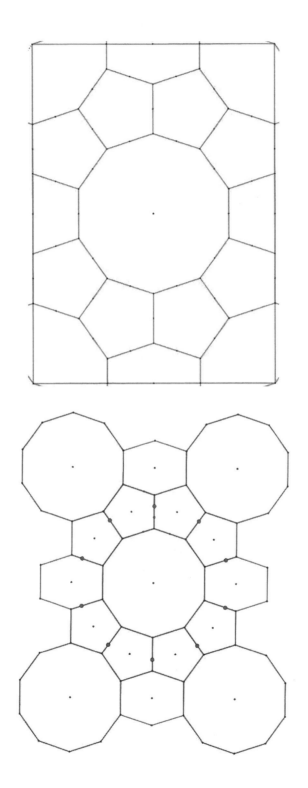

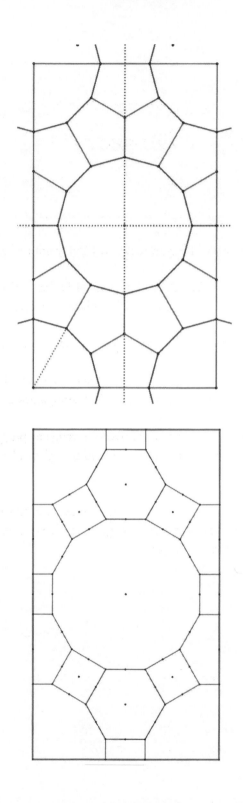

Glossary

Grid: A set of lines or line segments used to cover a region.

Plane: A two-dimensional region that may or may not be bounded.

Polygon-in-Contact (PIC) Method: A method of design that first creates an underlying grid of polygons upon which is placed the final design.

Polygon: A plane two-dimensional shape with straight edges.

Regular Tiling: A tiling composed of identical polygons, such as four squares, six triangles, or three hexagons, with identical configuration at each vertex.

Semi-Regular Tiling: A tiling composed of a range of polygons with every vertex identically arranged, e.g., hexagons and triangles or hexagons, squares, and triangles.

*Tiling (*or *Tessellation):* A covering of the plane with polygons, sometimes called tiles. In other words, it is an arrangement of shapes, such that they fill the plane without leaving any gaps.